Intensified Sediment Disasters in Japan

The 2011 Kii Peninsula disaster was postwar Japan's largest sediment and flood disaster. This book analyzes the disaster and the emergency response and subsequent disaster-prevention efforts. It also provides an international comparison and recommendations for mitigation and recovery efforts.

Although the scale and intensity of the disaster were expected to occur just once every 100 years, global warming has seen the intensification of such disasters around the globe. This book therefore presents an invaluable in-depth reference for readers on how to prepare for such a disaster, identify risk factors, and react accordingly. Contributors draw on the results of field surveys conducted by the Japanese Geotechnical Society at the time of the disaster and subsequent developments. First, they explain the factors that contributed to the disaster, including the meteorological, topographical, and geological conditions at the time of the disaster. They then describe the mechanisms of slope failure and damage caused by the slope failures across Nara, Wakayama, and Mie prefectures. Finally, they describe the post-disaster response, including the recovery and reconstruction and disaster-prevention and mitigation efforts in the affected area. Readers will therefore understand the importance of the contributing factors and be able to improve disaster mitigation strategies and response plans that will save lives and prevent damage to local infrastructure and economies.

This book is an invaluable resource for researchers, geologists, practicing engineers, and government officials who are involved in disaster prevention and response. Upper undergraduate and graduate students will also benefit from the book's in-depth approach.

Ryoichi Fukagawa, PhD, is a Professor at Ritsumeikan University, Kyoto, Japan. He earned a BE and an ME in civil engineering at Kyoto University in 1977 and 1979, respectively. He is a member of several societies, including the Japan Society of Civil Engineers, the Japanese Geotechnical Society, the International Society for Soil Mechanics and Geotechnical Engineering, and the International Society for Terrain-Vehicle Systems. He was the head of the Kii Peninsula Disaster Study Group, on which this publication is based, as well as the chairman of the subsequent Research and Study Committee.

Intensified Sediment Disasters in Japan

The 2011 Kii Peninsula Torrential Rain Disasters

Edited by Ryoichi Fukagawa

CRC Press
Taylor & Francis Group
Boca Raton London New York

CRC Press is an imprint of the
Taylor & Francis Group, an **informa** business

First edition published 2024
by CRC Press
4 Park Square, Milton Park, Abingdon, Oxon, OX14 4RN

and by CRC Press
2385 NW Executive Center Drive, Suite 320, Boca Raton FL 33431

British Library Cataloguing-in-Publication Data
A catalogue record for this book is available from the British Library

ISBN: 978-1-032-45067-4 (hbk)
ISBN: 978-1-032-45066-7 (pbk)
ISBN: 978-1-003-37521-0 (ebk)

DOI: 10.1201/9781003375210

Typeset in Times
by MPS Limited, Dehradun

Contents

Acknowledgments

The impetus for publishing this book came from an email sent by Andrew Stow of CRC Press. He asked if I would like to publish a book in English. Although I had co-authored and published several books in Japanese, I had no experience in English. I was a little apprehensive at first; however, after discussing the publication with Mr. Stow, I was almost convinced of the significance of publishing this book. The Kii Peninsula Disaster was a historic disaster of a scale that occurs only once every 100 years in Japan. We are convinced that describing this historic disaster will have important meaning and significance for sediment-disaster-prone countries; therefore, lessons can be learned for those countries. I would like to thank Mr. Stow for making the publication of this book possible. I would like to express my deepest gratitude to him. I would also like to thank Ms. Nivedita Menon and Mr. Mayank Sharma of CRC Press. They were kind enough to take charge of our book. The editing process of our book went smoothly because of their continuous encouragement.

We would like to thank many people who helped us in the preparation of this publication. As mentioned in the preface of this book, the motivation for writing this book was the survey conducted by the Joint Survey Team on Geo-Hazards in the Kii Peninsula Caused by Severe Tropical Storm Talas (1211). The survey team was made up of members of five organizations: the Japanese Geotechnical Society, the Geological Society of Japan, the Japan Society of Engineering Geology, the Kansai Geological Survey Association, and the Chubu Geological Survey Association. Because of the vastness of the affected area, the survey was conducted in the three prefectures involved, namely Nara, Wakayama, and Mie. The survey team consisted of 30 members for Nara, 33 for Wakayama, and 15 for Mie, a total of 78 members. In addition to the five named organizations, the Kinki Regional Development Bureau of the Ministry of Land, Infrastructure, Transport and Tourism; Nara, Wakayama, and Mie prefectures; and other related local governments have provided tremendous support and assistance in conducting field surveys and providing materials for the joint survey team. After the report of the above survey team was compiled, there was a strong movement to continue the field survey and analysis, and the members of the above five organizations almost immediately joined the Survey and Research Committee on Response to Geo-Hazards Caused by "Unexpected" Heavy Rainfall, which was established by the Kansai Branch of the Japan Geotechnical Society. It had 76 members. This committee has been actively engaged in research activities for three years starting in 2012 and compiled a report as the research committee in March 2015. There were 38 authors in charge of writing the report. A summary of this report was published serially as Series Lectures in the *Journal of Japan Geotechnical Society*. The authors are as follows. Persons marked with * in the table below are the authors of this book.

Chapter	Title	Authors	Affiliation
1	Introduction	Fukagawa, R.*	Ritsumeikan Univ.
2	Rainfall Characteristics of Typhoon No. 12 in 2011 and Topography and Geology of the Kii Peninsula	Mitamura, M.* Hirai, K.	Osaka Metropolitan Univ. Atec Yoshimura Co., Ltd
3	Disasters in the Nara Prefecture	Mitamura, M.* Torii, N. Hirai, K. Kagamihara, S.	Osaka Metropolitan Univ. Kobe City College of Technology Atec Yoshimura Co., Ltd Dia Consultants Co., Ltd
4	Disasters in the Wakayama Prefecture	Egusa, N. Yano, H. Tsujino, H. Nakanishi, N. Ishida, Y. Nabeshima, Y.*	Wakayama Univ. Chuo Kaihatsu Corp. Suncoh Consultants Co., Ltd Fukken Co., Ltd Kindai Univ. Technical College National Ins. of Tech., Akashi College
5	Disasters in the Mie Prefecture	Sakai, T. Okajima, K.* Konegawa, T. Ishikawa, M. Kataoka, Y. Sakaguchi, K.	Mie Univ. Mie Univ. Mihama Town, Town office Toho Chisui Co., Ltd Kinki Geo-Eng. Center Co., Ltd Asia Air Survey Co., Ltd
6	Disaster Prevention and Mitigation Efforts after the Kii Peninsula Disasters	Hioki, K.* Kobayashi, T.* Ushiro, S.* Okajima, K.* Koizumi, K.* Izunami, R.*	Osaka Ins. of Tech. Ritsumeikan Univ. Wakayama Univ. Mie Univ. Osaka Univ. West Japan Railway Company
7	Response to Ground Disasters Caused by Unexpected Heavy Rainfall and Lessons Learned	Fukagawa, R.* Tsukahara, J.	Ritsumeikan Univ. Chuo Kaihatsu Corp.
8	Conclusions	Fukagawa, R.*	Ritsumeikan Univ.

The members of the joint survey team and the subsequent survey and research committee came from a wide range of fields, including civil engineering, geotechnical engineering, geology, agricultural engineering, erosion control, and public administration, making the survey and analysis work a very stimulating setting for interdisciplinary exchange. I would like to express my profound gratitude to all the members who gathered here. In particular, this project would not have been possible without the help of Prof. Mitamura (Osaka Metropolitan University), Prof. Egusa (Wakayama University), and Prof. Sakai (Mie University), who organized the surveys in each prefecture in the joint survey team. I would like to express my gratitude to them. I would also like to thank Prof. Mitamura (Osaka Metropolitan University), Prof. Nabeshima (National Institute of Technology, Akashi College), Prof. Okajima (Mie University), and Prof. Hioki (Osaka Institute of Technology) as principal contributors to the chapters in this book. Although this book is based on a manuscript for "Series Lectures" published in the *Journal of the Japan Geotechnical Society,* many parts have been newly written. In this sense, we owe much to these four authors. I would like to express my gratitude to the above four authors for the collaboration that resulted in this book.

In February 2023, as I was finishing a draft of this book, a huge earthquake hit Turkey and Syria. After the disaster, the number of victims continued to rise daily, which was heartbreaking. According to reports since March, the victims exceeded 50,000, a catastrophic number. This book has mainly reported on sediment disasters, but there is concern that the synergistic effects of extreme weather events damaging the land and then a huge earthquake following on that damaged land will increase even more. In Japan, the probability of the Nankai Trough earthquake, a massive earthquake of the magnitude 9.0 class, occurring within the next 30 years is predicted to be around 80%. I write this while expressing my condolences to the victims of the Turkey-Syria earthquake and praying for the speedy recovery and reconstruction of the affected areas.

Ryoichi Fukagawa
March 2023

Preface

Japan is a nation of natural disasters. Earthquakes, typhoons, torrential rains, heavy snowfalls, and volcanic eruptions are among the many causes. Our ancestors suffered and endured these disasters, and somehow managed to overcome them. It is no exaggeration to say that we Japanese are still struggling against natural disasters on a daily basis. On the other hand, it may be said that we are insensitive to environmental changes caused by global warming. It may be that because there are so many natural disasters, it is difficult to notice the increase in such disasters caused by global warming.

The year 2011 was an unforgettable year for the Japanese people. More than 20,000 people were killed by the 9.0 magnitude earthquake that struck off the Pacific coast of Tohoku on March 11, 2011. The Kii Peninsula Disaster, the subject of this book, occurred in September of the same year and claimed nearly 100 victims, making it a once-in-a-century disaster in terms of the scale of the rainfall and the amount of collapsed sediment. From 2011 until 2021, large-scale sediment and flood disasters occurred every year, each claiming dozens of victims. We believe that Japanese people themselves should pay more attention to the abnormality of this situation. Disasters are so familiar to the Japanese that they are not perceived as unusual. Behind the scenes, anomalies, such as the first time in the history of observation something has taken place, have been occurring frequently. An example is the location of where typhoons hit land. Typhoons that hit Japan usually make landfall in Kyushu, Shikoku, and the southern part of Honshu from the Pacific Ocean or the South China Sea. However, extremely unusual conditions emerged in August 2016, with Tropical Storm Chanthu (1607) and Tropical Storm Kompasu (1611) making landfall in Hokkaido directly from the Pacific Ocean, and Typhoon Lionrock (1610) in August 2016 making landfall in Iwate Prefecture in the northern Tohoku region from the Pacific Ocean. In addition, in the torrential rains that have occurred since 2011, abnormal amounts of rain have been observed at many locations in various regions, the most in recorded history. It is normal to think that some kind of change in weather conditions is taking place.

Here, I would like to discuss the notation of typhoons in this book. According to the Japan Meteorological Agency's definition, a storm is defined as a "typhoon" if its maximum wind speed, which is the maximum value of the average wind speed over a 10-minute period, is 17 m/sec or higher. They are numbered according to the order of occurrence since the beginning of the year. On the other hand, international standards define typhoons differently. A "tropical storm" is defined as a storm with a maximum wind speed of 17–24 m/sec, a "severe tropical storm" as a storm with a maximum wind speed of 24–32 m/sec, and a "typhoon" as a storm with a maximum wind speed of 32 m/sec or higher. Typhoon No. 12 of 2011 or Typhoon (1112), the main subject of this publication, is referred to as "Severe Tropical Storm Talas" by international standards. In this publication, all typhoon designations follow the international standard.

Let me also explain a few technical terms related to sediment disasters. Sediment disasters are diverse. In this book, technical terms such as *large-scale slope failure, debris flow,* and *landslide* are frequently used. However, it is difficult to strictly unify these terms, to the extent that the definitions and terminology of these terms may differ slightly from one academic society to another. Many disaster-related terms used in newspaper reports and other media are not recognized as academic terms. The terms used in this book are those in standard use in academic societies. For simple definitions, please refer to Section 7.2.1 first.

This book was written under the circumstances described above. Although this book is written primarily for people outside of Japan, I hope that Japanese people will read it as well, in the sense that the effects of global warming should be taken a little more seriously.

Although the main theme of this book is sediment disasters, it also devotes many pages to flood disasters. However, the book does not focus on the hydrological aspects of flood disasters, such as flood flows, but mainly on the destruction of riverbanks and levees, and the destruction of bridges over rivers. In this sense, it can be said that this book is mainly concerned with sediment disasters.

The author specializes in geotechnical hazards. When analyzing slope failure phenomena, at a minimum, rainfall conditions, topographic and geologic conditions of the slope of interest, and the time of occurrence of the failure must be known. Merely describing collapse phenomena does not help with the challenge of predicting slope failure phenomena. The occurrence of the last collapse time is the most difficult factor to identify. For large-scale slope failures, methods have been developed to estimate the location and time of occurrence based on information from seismic intensity gauges installed throughout Japan, but for small-scale slope failures, the time of occurrence is often unknown. The topography and geology of the slope are not easy to ascertain. It is common for one slope to collapse while an adjacent one avoids collapse. However, it is often difficult to explain why. As mentioned, looking at sediment disasters analytically is not an easy task. This publication is the product of many members' struggles with sediment disasters. We sincerely hope that it will be informative and useful to many of you.

Ryoichi Fukagawa
Ritsumeikan University

.

1 Introduction

Ryoichi Fukagawa

1.1 Motivation

The motivation for writing this book is that all of the authors experienced the Kii Peninsula Disaster described below and were involved in the investigation and analysis of the disaster after it occurred. We felt that this was truly a historic disaster, one that should be passed on to future generations.

1.2 Kii Peninsula Disasters

The Severe Tropical Storm Talas (1112) moved slowly northward on a path through the eastern part of Shikoku Island from August 30 to September 4 in 2011. As the result of Severe Tropical Storm Talas, heavy rains continued for a long period of time, especially in the Kii Peninsula, causing extensive damage mainly in Nara, Wakayama, and Mie prefectures. In the three prefectures, 88 people were killed or missing, 369 houses were completely destroyed, 2,901 houses were partially destroyed, and 7,531 houses were flooded (including those above and below floor level). [1]

The Kii Peninsula has been hit by large-scale torrential rain disasters many times in the past. Japan has a unique chronology for the reign of its emperors. In modern Japan, they are Meiji (1868–1912), Taisho (1912–1926), Showa (1926–1989), Heisei (1989–2019), and Reiwa (2019–). The Kii Peninsula experienced large-scale sediment and flood disasters in the Meiji, Showa, and Heisei eras. All of these disasters were of historical proportions. The main subject of this book is the 2011 Heisei Kii Peninsula Torrential Rain Disaster, which was preceded by the Showa Kii Peninsula Torrential Rain Disaster in July 1953 and the Meiji Kii Peninsula Torrential Rain Disaster, known as "Totsu River Flood Disaster," in August 1889.

Among these three disasters, the Meiji Kii Peninsula Torrential Rain Disaster caused extensive damage to the Kii Peninsula. A typhoon hit the area on August 19–20, 1889, resulting in torrential rains with a cumulative rainfall of more than 1,000 mm [2]. As a result, many large-scale slope failures occurred from the Totsu River Basin upstream of the Kumano River in Nara Prefecture to the Hidaka River and Tomita River Basins in Wakayama

DOI: 10.1201/9781003375210-1

Prefecture, forming and breaking river channel blockages in many areas. These sediment and flood disasters caused extensive damage. The number of dead or missing was 1,513, and the number of houses washed away or destroyed completely rose to 3,811 and 1,551 in Nara and Wakayama prefectures, respectively [2]. Among the sediment transport phenomena that formed river channel blockages, there were five cases in which the amount of transported sediment exceeded 10 million m^3, and the entire Totsu River Basin exceeded 100 million m^3 [3].

The Showa Kii Peninsula Torrential Rain Disaster was also a record-breaking disaster. The torrential rains caused by the rainy season front that hit the northern part of Wakayama Prefecture from July 17 to the morning of July 18 in 1953 brought record rainfall, including cumulative rainfall of 1,500 mm and hourly rainfall of 80–100 mm observed in Hanazono Village, Wakayama Prefecture. It caused extensive damage, with 1,066 people dead or missing, 4,407 houses washed away, and 4,193 houses destroyed completely. Although this disaster is sometimes referred to as the "Arita River Flood," it is important to note that the damage was not limited to the Arita River Basin, as the Kishi River, Hidaka River, and Kumano River Basins were also severely affected [4].

Photo 1.1 [4] shows monuments in Hongu Town, Tanabe City, Wakayama Prefecture, that indicate the maximum water levels of the three Kii Peninsula flood disasters. It clearly shows the magnitude of the Meiji Kii Peninsula Disaster. Although the damage to people and property of the 2011 Heisei Disaster was far less than those of the Meiji and Showa Disasters, the amount of rainfall and the scale of floods and slope failures that occurred were almost comparable to those of these historic disasters.

1.3 Research and Analysis of Kii Peninsula Disaster

Anyway, the main target of this book is the 2011 Kii Peninsula Disaster. The Kansai Branch of the Japan Society of Geotechnical Engineers (JGSE) established the Joint Survey Team on Geotechnical Hazards Caused by the Severe Tropical Storm Talas to investigate the above-mentioned disasters. Because the damage was extensive and widespread, and because the rainfall, topography, and geology were expected to be complex, the project was conducted jointly with the Geological Society of Japan, the Japan Society of Applied Geology, the Kansai Association of Geological Surveyors, and the Chubu Association of Geological Surveyors. The survey was conducted on a prefectural basis, with 30 members of the Nara Group, 33 members of the Wakayama Group, and 15 members of the Mie Group. The group leaders were, in order, Prof. Mitamura (Osaka City University), Prof. Egusa (Wakayama University), and Prof. Sakai (Mie University), and the survey team was headed by Prof. Fukagawa.

The Research and Study Committee on Response to Ground Hazards Caused by "Unexpected" Heavy Rainfall was formed on the basis of the

Photo 1.1 Monuments in Hongu Town, Tanabe City, Wakayama Prefecture, indicating the maximum water levels of the three Kii Peninsula flood disasters [4].

above joint research team. It was jointly operated by four academic societies: the Kansai Branch of the Geotechnical Engineering Society of Japan, the Japan Society of Applied Geology, the Kansai Association of Geological Surveyors, and the Chubu Association of Geological Surveyors. Since its establishment, the Kansai Section has been actively engaged in surveys and research activities aimed at understanding the actual damage, elucidating disaster mechanisms, and proposing evacuation and disaster-prevention measures. The committee was implemented as a three-year plan starting in FY2012 with the participation of about 70 committee members. The "Series Lecture" in the Journal of Geotechnical Engineering, which was the basis for this publication, summarizes the results of this committee's activities [5–13]. The term "unexpected" is included in the committee's name for the following reasons. Terms such as "unexpected" or "unprecedented" should not be used carelessly in official documents, but in practical situations such as concrete design, it is sometimes necessary to assume external forces that exceed the level of those normally used. With such a situation in mind, and with a sense of self-discipline, the term "unexpected" is used with " ".

1.4 Purpose of This Book

Large-scale sediment disasters have been occurring one after another around the world. The trend toward more frequent, widespread, and more severe sediment disasters is expected to intensify in the future, partly due to the advance of global warming. Japan is, of course, no exception, and the trends described above have become increasingly pronounced in recent years. The 2011 Kii Peninsula Disaster was a historic sediment disasters that occurs only once in a century in Japan. The authors spent several years after the disaster conducting field surveys and analyzing the results. Therefore, the purpose of this book is to provide information that will be useful for disaster prevention and mitigation in the affected areas, as well as for people around the world suffering from similar sediment disasters.

1.5 Outline of This Book

The main part of this book consists of eight chapters. The principal contributors of each chapter are as follows; Chapters 1, 3, 7, and 9: Prof. Fukagawa of Ritsumeikan University; Chapters 3 and 4: Prof. Mitamura of Osaka Metropolitan University; Chapter 5: Prof. Nabeshima of National Institute of Technology, Akashi College; Chapter 6: Prof. Okajima of Mie University; and Chapter 8: Prof. Hioki of Osaka Sangyo University. This book is the result of collaboration with the above researchers, although only my name appears on the cover of the book as editor.

The chapters are summarized as follows.

Chapter 1 begins by describing the circumstances that led to the writing of this book. Next, it provides an overview of the book.

Chapter 2 first summarizes the recent occurrence of sediment disasters in the world. It shows that while in the past sediment disasters tended to occur more frequently in certain countries in Asia and Central and South America, in recent years this trend has been expanding to Africa, Europe, and other regions. Next, we summarize the recent occurrence of sediment disasters in Japan. The report shows that torrential rains have become more frequent, more widespread, and more severe, and that large-scale sediment disasters occur every year as a result.

Chapter 3 describes the rainfall characteristics observed in the Kii Peninsula with the Severe Tropical Storm Talas. The cumulative rainfall exceeded 1,000 mm at many locations on the Kii Peninsula, including Kamikitayama Village in Nara Prefecture, where the 3-day rainfall reached 1,653.5 mm, the highest 3-day rainfall in recorded history in Japan. As an overview of the rainfall characteristics, the coastal areas experienced heavy rainfall reaching

100 mm/hour from midnight on September 3 to dawn on September 4, while the mountain areas experienced heavy rainfall of 30–40 mm/hour for a long time, resulting in a cumulative rainfall of over 1,000 mm. These differences in rainfall characteristics have a significant influence on the mechanism of slope failure in both regions. Chapter 3 also describes the geomorphological characteristics of the Kii Peninsula. The geomorphological conditions in the Kii Peninsula are basically regulated by accretionary complexes. The Kumano Group sedimentary rocks overlie some of the accretionary complexes, and the Kumano acidic igneous rocks intrude the accretionary complexes and the Kumano Group. These geomorphological conditions also have a significant influence on the mechanism of slope failure.

Chapter 4 describes the actual situation of disasters in the Nara Prefecture. The most significant feature of disasters in Nara is the predominance of large-scale slope failures ("deep-seated failures"). Large-scale slope failures occurred in many areas, such as Akadani, Nagatono, and Ui, which led to the formation of natural dams, and placed tension on the local communities and disaster management agencies for a long period of time after the collapse. This chapter analyzes the mechanisms of these large-scale slope failures based on the results of detailed field investigations. On the other hand, surface failures also occurred frequently in Nara Prefecture. In addition to the analysis of the factors causing these surface failures, field investigations, estimation of ground properties based on sampled soils, and slope stability analysis of a slope failure site in Nosegawa Village are attempted, showing that the slope failure mechanism can be expressed quantitatively to some extent.

Chapter 5 describes the actual situation of disasters in Wakayama Prefecture. The characteristic of disasters in Wakayama is that large-scale debris flows were more frequent than large-scale slope failures. So-called slope surface failures also occurred frequently. Large-scale debris flows, especially in Shingu City and Nachikatsuura Town, occurred mostly on slopes composed of Kumano acidic igneous rocks. The occurrence time was concentrated at the dawn of September 4, which may have been triggered by the intense rainfall exceeding 100 mm per hour observed during this period. The effects of these factors on slope surface failures are analyzed based on much data. Wakayama is also characterized by a large number of river disasters. This chapter reviews them and discusses their causes.

Chapter 6 describes the actual situation of disasters in Mie Prefecture. The characteristics of disasters in Mie are similar to those in Wakayama. That is, large-scale slope failures occur in mountainous areas, and surface slope failures and large-scale debris flows increase as the coastal areas are approached. Large-scale slope failures are basically caused by accretionary structures, and fortunately, in the case of Mie Prefecture, the impact on the region was not so great. The study also includes a detailed examination of surface slope failures, and a large amount of data shows that the threshold for the occurrence of surface failures is 800 mm cumulative rainfall and 80 mm hourly rainfall. River disasters are also especially severe in Kiho Town adjacent to Shingu City, where a relatively recently constructed levee failed, prompting a reconsideration of future countermeasures.

Chapter 7 provides an overview of slope disaster prevention measure works in Japan. Measure works should be planned and constructed by comprehensively considering the topography and geological conditions of the slope, anticipated slope hazards, and the local rainfall environment. The anticipated slope hazards are steep slope failure, landslide, debris flow, and rock fall. Measure works can be roughly classified into two categories: preventive works, which are often implemented as emergency measures, and protective works, which are implemented as permanent measures. This chapter introduces typical measure works in order. This chapter also introduces representative measure works actually constructed in the 2011 Kii Peninsula Disaster in Nara, Wakayama, and Mie prefectures, the main subject of this report.

Chapter 8 introduces the efforts for disaster prevention and mitigation after the Kii Peninsula Disasters. The committee has developed a risk prediction and monitoring system for deep-seated landslide during heavy rainfall based on a modified tank model, mainly for the mountainous areas in Nara Prefecture, and is still conducting risk assessment. This chapter further introduces a collection of initiatives that have been initiated in response to the recent flooding. Examples include flood and inundation countermeasures in Shingu City and Nachikatsuura Town, a timeline in Kiho Town, and efforts by road and railroad operators.

Chapter 9 summarizes recommendations for governments, facility managers, and the general public in the hope of reducing the damage caused by large-scale heavy rainfall disasters that are expected to occur in the future.

References

[1] Cabinet Office: Damage caused by Severe Tropical Storm Talas (1112), 2011.9.28. (in Japanese)
[2] Tada, Y., Daimaru, H. and Sammori, T.: Characteristics and differences of flood disaster through all times in Kinki area, Japan, Sabo, Vol.65, No.5, pp.58–64, 2013. (in Japanese)
[3] Based on construction record provided by the Sabo Office of Kii Mountain Range, 2023. 3. (in Japanese)
[4] Ushiro, S.: Kii Peninsula Great Torrential Disaster, Haru-Shobo Publishing Co. Ltd., 2022. 10. (in Japanese)
[5] Edited by four organizations including Kansai Branch of the Japan Society of Geotechnical Engineers (JGSE): Report of the Research and Study Committee on Response to Ground Disasters Caused by "Unexpected" Heavy Rainfall, 2015. (in Japanese)
[6] Fukagawa, R.: The 2011 Kii Peninsula flood and sediment disasters: facts and lessons learned, Journal of the Japan Society of Geotechnical Engineers (JGSE), Introduction, "Series Lecture", No.1, pp.43–44, 2016. 4. (in Japanese)

[7] Mitamura, M. and Hirai, K.: ditto, "Series Lecture", No.2, Summary of rainfall characteristics and topography and geology of Kii Peninsula by Typhoon No. 12 in 2011, pp.45–52, 2016.4. (in Japanese)

[8] Mitamura, M., Torii, N., Hirai, K. and Kagamihara, S.: ditto, "Series Lecture", No.3, Disasters in Nara Prefecture, pp. 40–47, 2016.5. (in Japanese)

[9] Egusa, N., Yano, H., Tsujino, H., Nakanishi, N., Ishida, Y. and Nabeshima, Y.: ditto, "Series Lecture", No.4 Disasters in Wakayama Prefecture, pp.41–48, 2016.6. (in Japanese)

[10] Sakai, T., Okajima, K., Konegawa, T., Ishikawa, M., Kataoka, Y., and Sakaguchi, K.: ditto, "Series Lecture", No.5, Disasters in Mie Prefecture, pp.50–57, 2016.7. (in Japanese)

[11] Hioki, K., Kobayashi, T., Ushiro, S., Okajima, K., Koizumi, K., and Izunami, R.: ditto, "Series Lecture", No.6, Disaster prevention and mitigation efforts after the Kii Peninsula Disaster, pp.69–75, 2016.8. (in Japanese)

[12] Fukagawa, R. and Tsukahara, J.: ditto, "Series Lecture", No.7, Response to ground disasters caused by unexpected heavy rainfall and lessons learned, pp.40–45, 2016.9. (in Japanese)

[13] Fukagawa, R.: ditto, "Series Lecture", No.8, Conclusions, p.46, 2016.9. (in Japanese)

2 Intensification of Sediment Disasters in the World and in Japan

Ryoichi Fukagawa

2.1 What Happens in the World?

2.1.1 Global Warming and Extreme Weather

Global warming is accelerating. In 2021, Canada experienced a maximum temperature of 49.6°C and a heat wave that killed 233 people (June) [1]. In Germany and Belgium, river flooding and sediment disasters killed more than 200 people (July) [2]. Mediterranean countries such as Italy, France, and Greece were hit by a heat wave of nearly 50°C and massive wildfires (August) [3]. Hurricane-related flooding in the northeastern United States, including New York City, killed 41 people (September) [4]. In 2022, 8 people were killed and 15 others went missing when a glacier collapsed in northern Italy (July) [5]. Europe was hit by an intense heat wave. Temperatures exceeded 40°C in England, and more than 1,000 people died in Portugal and Spain (July) [6]. Pakistan was hit by prolonged heavy rains that inundated one-third of the country, killing 1,162 people and affecting more than 33 million people (from June) [7]. The list goes on and on. In recent years, Europe, which had been thought to be relatively free from natural disasters, has been hit by a number of them.

Many researchers and journalists have sounded the alarm about global warming. However, since global warming is complicated by many factors, most researchers have been rather cautious in their assessment of the causal relationship between global warming and human activities. A recent book that warns about the progress of global warming is *The Uninhabitable Earth Life after Warming* [8] by David Wallace-Wells. The book was published in 2019. Looking at the various events that occurred in 2021 and 2022, Wells seems to be correct in his assertion. In addition, David Attenborough, known for his excellent natural history documentaries, expresses serious concern about the progress of global warming in his recent book, *A Life on Our Planet* [9]. He asserts that if we do not address "green growth" and "rewilding" now, there will be no future for humankind and the planet.

Figure 2.1 [10] shows the evolution of annual average temperatures worldwide with respect to the average for the period 1991–2020. The dark solid blue line in the figure represents the moving average, and the red straight line

DOI: 10.1201/9781003375210-2

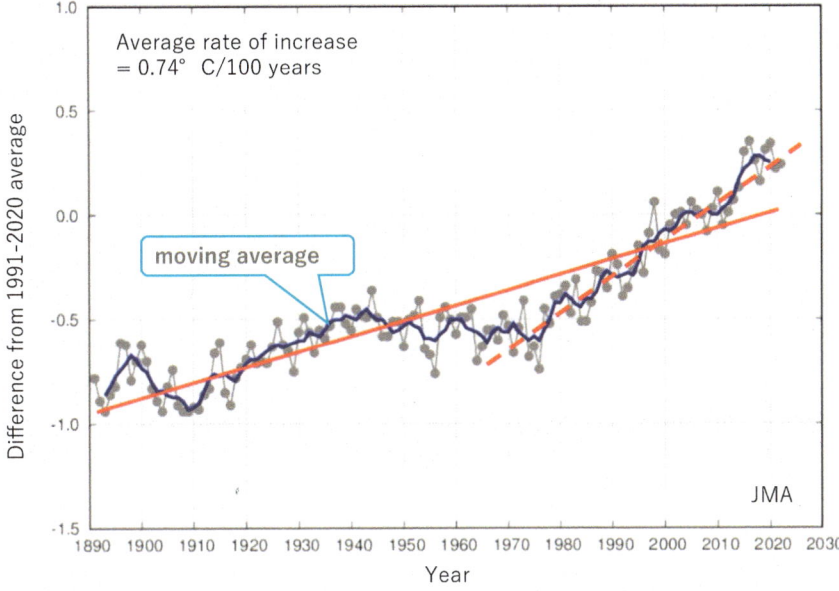

Figure 2.1 Change in global average annual temperatures with respect to the 1991–2020 average [10].

represents the least-squares approximation, indicating that the trend of rising temperatures has become even stronger since 1980. Against this background, the first working group (WG1) of the United Nations Intergovernmental Panel on Climate Change (IPCC) reported in August 2021 [11] that even if countries take the best possible measures to combat global warming, the global average temperature will rise 1.5°C above pre-industrial levels in the next 20 years. For the first time, the IPCC also determined that the temperature increase is caused by humans. The report was prepared by more than 750 experts from around the world based on the latest research findings and surveys, so it should be considered the consensus of the entire scientific community. This fact must be taken seriously. We should also be more imaginative about what a 1.5°C increase in average temperatures will bring. Heat waves will become more frequent. More wildfires. More people will starve. More countries will be submerged because of rising sea levels. Typhoons will intensify and become larger. Droughts will become more frequent. Infectious diseases will spread. And so on. Data on the causal relationship between rising temperatures and the above results are accumulating one after another. It is also important to note that temperature rise will not occur uniformly across the globe. There is a risk that harsher regions will emerge and become "uninhabitable."

2.1.2 Sediment Disasters in the World

The main subject of this publication is sediment disasters. Sediment disasters are often smaller in scale than big flood or earthquake disasters and therefore rarely make the international news. For this reason, it has not been easy to understand the recent trend in sediment disasters around the world. However, with the construction of various databases on natural disasters, it has become possible to obtain a complete picture. For example, the Sabo and Landslide Technical Center (STC) in Japan has compiled a chronological summary of the occurrence of sediment disasters worldwide since 2010 [12]. The main data sources are listed below.

EM-DAT [13], Relief Web [14], Reuter Alertnet [15], AFPBB Disaster News [16], NHK OnLine [17].

The sediment disasters include slope failures, debris flows, and landslides, although STC also classifies pyroclastic flows caused by volcanic eruptions, and so on, which are not included in this analysis. Disasters with ten or more fatalities or missing people are included in this analysis. Since sediment disasters are more localized than flood disasters, if more than ten people are killed in a single sediment disaster, that indicates it is a large-scale sediment disaster. Please refer to the original source [12] for the procedure of data extraction and the accuracy of the extracted data.

First, let's look at the number of large-scale sediment disasters in Asia, Africa, North America, Central and South America, Europe, the Middle East, and Oceania (Figure 2.2). It is clear that the Asia region has by far the largest number of sediment disasters. On average, Asia experiences about 17 major sediment disasters per year. This is because many of the countries in the region have tropical rainforest climates or warm and humid climates,

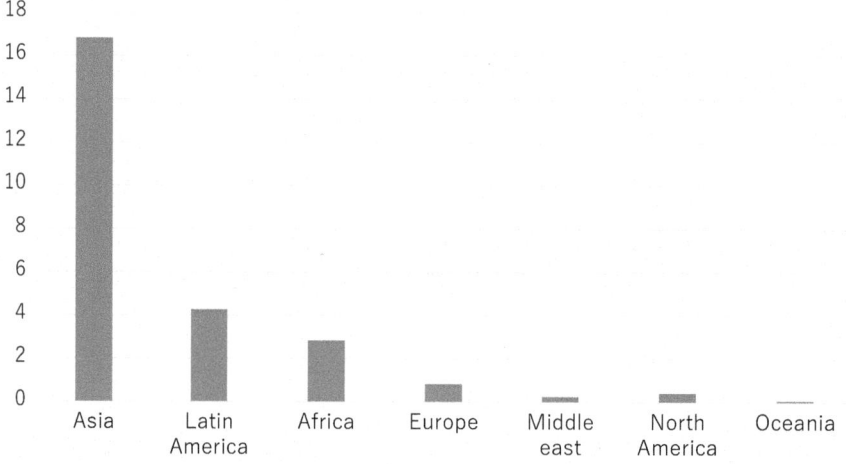

Figure 2.2 Average number of large-scale sediment disasters per year during 2010–2021.

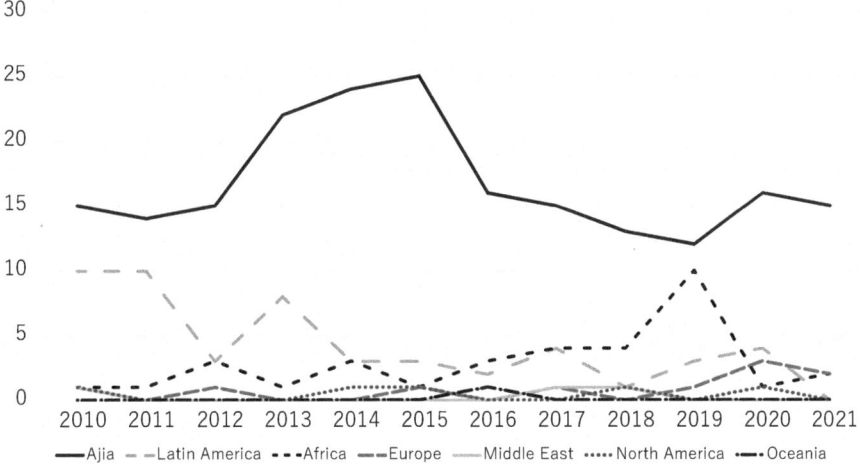

Figure 2.3 Change over the years of large-scale sediment disasters per year during 2010–2021.

generally with high rainfall, and many of the countries are predominantly mountainous due to plate tectonics. In Europe, sediment disasters have been considered rare, but their number has been increasing, as evidenced by the fact that they now occur almost once a year. The results show that large-scale sediment disasters are less likely to occur in North America, the Middle East, and Oceania.

Next, let's take a look at the number of large-scale sediment disasters in each region over time. The results are shown in Figure 2.3, which indicates that Asia experiences an average of 15–20 major sediment disasters per year, reflecting the high number of sediment disasters from 2013 to 2015, especially in Indonesia, China, Nepal, and India. Central and South America had a particularly high number of sediment disaster events in 2010, 2011, and 2013. This is due to the occurrence of disasters in countries such as Brazil, Mexico, and Colombia. Since then, the number of sediment disasters has settled at about three per year. By region, the trend toward more frequent sediment disasters in Africa and Europe is worrisome. As mentioned earlier, the occurrence of flooding and sediment disasters in Germany and Belgium is symbolic of this trend.

By the way, Figure 2.4 [10] shows the global precipitation trend based on the average annual precipitation from 1991 to 2020. The negative side indicates drought conditions, while the positive side indicates more precipitation. The data clearly show drier conditions up to about 2000, with some drier years after 2000, but also more years with more precipitation. Originally, it was said that the effect of global warming on weather would be that wet years would lead to more rain, and conversely, dry years would lead to severe droughts. Recent extreme weather events around the world seem to indicate

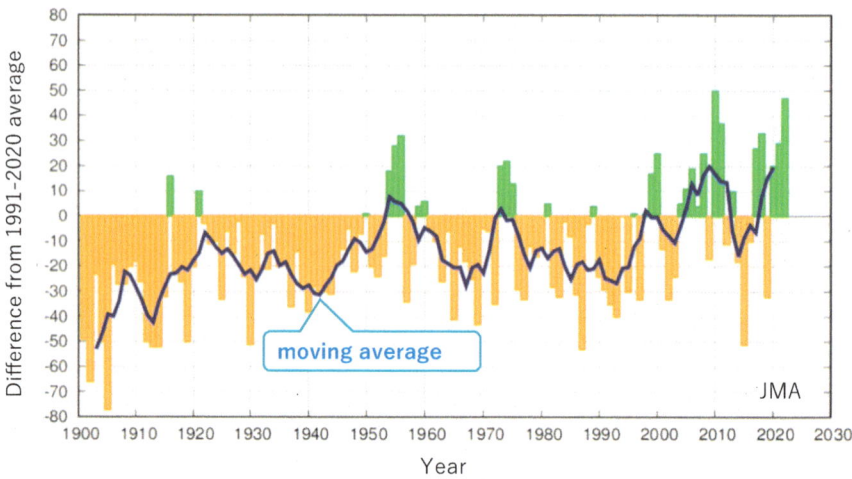

Figure 2.4 Change in global average annual precipitation with respect to the 1991–2020 average [10].

that this trend is becoming more and more pronounced. It is difficult to determine from Figure 2.4 whether there is a close relationship between global warming and large-scale sediment disasters in the world, but it is possible that the current trend of increasing sediment disasters in Africa and Europe can be attributed to global warming.

Finally, let's look at the incidence of sediment disasters by country. Figure 2.5 shows the ten countries with the highest number of incidents from 2010 to 2021, listed in descending order of frequency. A total of eight of the ten countries are in Asia. The remaining two countries are in South America. The largest number of incidents took place in Indonesia, which has a tropical rainforest climate and is close to a plate boundary, resulting in active crustal movement and severe topographical and geological conditions. The results reflect these conditions. China was the next most affected country. This reflects the fact that China is a large country with many places that have topographical, geological, and rainfall conditions that can easily cause sediment disasters. India, the Philippines, and Nepal had an average of about two large-scale sediment disasters per year, followed by Brazil, Colombia, Japan, and Viet Nam, which had an average of about one large-scale sediment disaster per year.

The above is a review of the occurrence of sediment disasters around the world. Compared with earthquake disasters and river flooding, slope failure events are often limited in terms of damage because of the small area affected locally. However, there have been some huge sediment disasters that have killed more than 100 people. The following is a brief introduction to the 2009 debris flow disaster in Xiaolin Village, Taiwan; the 2010 debris

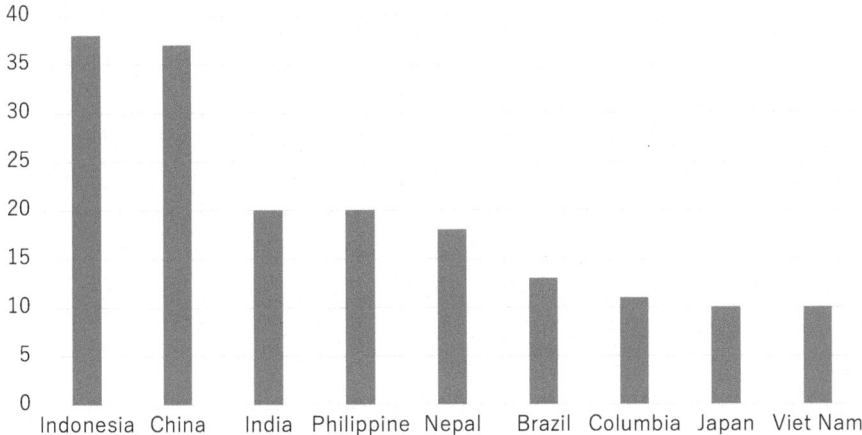

Figure 2.5 Total number of large-scale sediment disasters during 2010–2021.

flow disaster in Zhouqu County, China; and the 2012 Philippine Typhoon Bopha disaster.

2.1.3 2009 Debris Flow Disaster in Xiaolin Village, Taiwan [18]

From August 7 to 11, 2009, Typhoon Morakot approached Taiwan. This typhoon developed from a tropical depression on August 4, and moved westward with increasing strength, recording a central pressure of 940 hPa and maximum wind speeds of 80 m/sec just before making landfall on Taiwan. The typhoon slowed down from August 7 to 9, exerting a prolonged influence on Taiwan. Rainfall from the typhoon was concentrated in central and southern Taiwan, and the Alishan Observatory in central Taiwan recorded a cumulative rainfall of 3,022.5 mm from August 7 to 11 (the highest on record) and a daily rainfall of 1,165.5 mm (on August 9).

Xiaolin Village is located on the left terrace of Qishan Creek, a branch of Gaoping Creek. The steps of the Xiaolin Village tragedy are shown in Figure 2.6. At around 6 a.m., 190 hectares of mountainside behind Xiaolin Village collapsed on August 9, and the collapsed sediment covered the northern half of the village. This collapsed sediment blocked the Qishan Creek and formed a natural dam, which broke at around 7:00 a.m. and swept away the southern half of Xiaolin Village. By the time the natural dam was formed, the Qishan Creek tributary (a stream on the downstream side of the village) had already been washed away by the debris flow, causing the No. 8 and No. 9 bridges downstream of Xiaolin Village to be washed away, thus cutting off traffic and making it difficult to evacuate to the outside of the village. With the exception of residents who evacuated to the mountains, most residents lost their lives. The total number of dead or missing was 453, making it a catastrophic disaster.

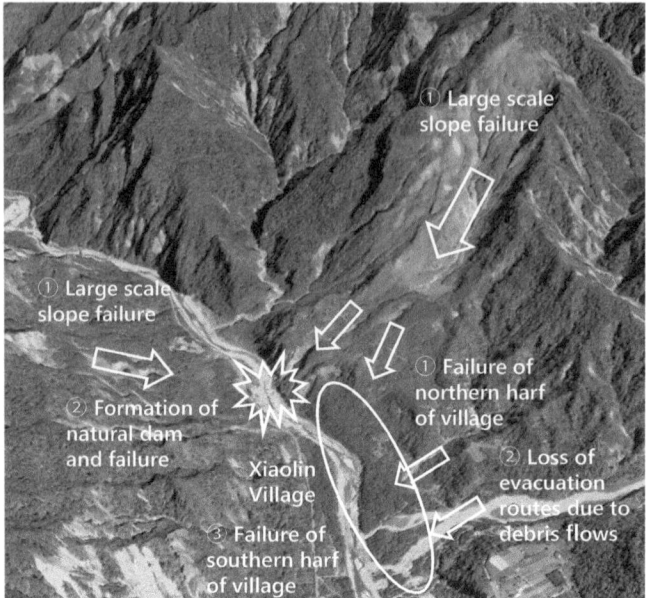

Figure 2.6 Steps of disaster at Xiaolin Village, Taiwan (Created by adding to Google map).

2.1.4 2010 Debris Flow Disaster in Zhouqu County, China [19]

In the early hours of August 7, 2010, a massive debris flow caused by torrential rains in Zhouqu County, Gannan Tibetan Autonomous Prefecture, Gansu Province, China, caused extensive damage. The rivers involved were the Sanyanyu and the Luojiayu. The average annual precipitation around this basin is 435.6 mm, and the area has always had low precipitation. On the other hand, debris flows are common in this area, occurring when the 10-minute rainfall intensity is 5 to 8 mm or more and the 30-minute rainfall intensity is 10 mm or more. Relatively large debris flows occur on average two or three times a year. The geological structure of the area is very complex and has well-developed faults from the influence of orogenic movements in the Yanshan Mountains of China and the Himalayas, as well as frequent earthquakes in the area. This is one of the reasons for the high incidence of sediment disasters.

The debris flow occurred at around 11:30 p.m. on August 7 and was 500 m wide and 5 km long. The rainfall that caused the debris flow was 77.3 mm/hour from 11 p.m. to midnight and 99.6 mm/hour in 4 hours. Although this is a small amount of rainfall compared with areas that typically have heavy rainfall, it is a different story where rainfall is usually low. The amount of sediment discharged by the debris flow amounted to approximately 1,420,000 m^3 at the Sanyanyu River and 330,000 m^3 at the Luojiayu River. The debris flows hit the

Photo 2.1 The city area swallowed by huge debris flow at Zhouqu County, China [19].

urban areas downstream, leaving 1,765 people dead or missing. Photo 2.1 [19] shows the city swallowed by the huge debris flow.

2.1.5 *2012 Philippine Typhoon Bopha Disaster [20–22]*

Typhoon Bopha made landfall in southern Mindanao on December 4, 2012. Classified as Category 5 by U.S. weather experts, Bopha crossed Mindanao

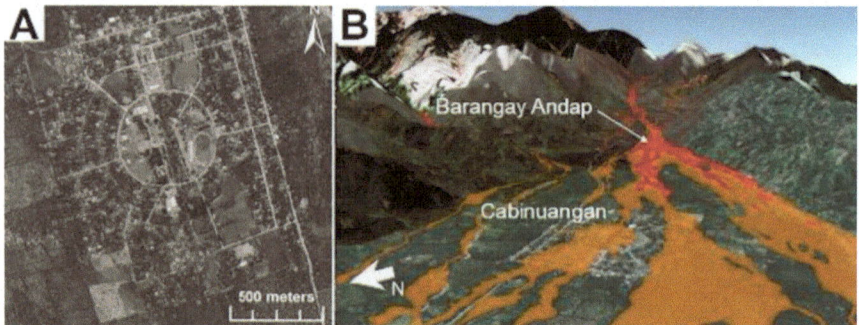

Figure 2.7 New Bataan. A: Andap, Google image of Cabinuangan (the central district of New Bataan) before the debris flow. B: Southward facing three-dimensional terrain diagram of Andap and Cabinuanga after the Mayo River disaster. Red areas are boulder-rich "true debris" flow; orange areas are deposits of more dilute "hyperconcentrated" flows [22].

with maximum winds of 49 m/s through Davao Oriental, Compostela Valley, Agusan del Sur, Bukidnon, and Misamis Oriental. It then continued west-northwest through Negros Oriental, and the Sulu Sea, Palawan Island and reached the West Philippine Sea before reversing course to northern Luzon Island. It disappeared before coming ashore.

With the passage of Bopha, torrential rains caused frequent slope failures and debris flows in many areas. The storm caused extensive damage to Mindanao, leaving 1,067 dead, 834 missing, and 2,666 injured islandwide, and displaced 970,000 people in 220,000 households, affecting 6.24 million people in 710,000 households. The worst affected areas were the southeastern region of Dabao Region, where a massive debris flow occurred. In Barangay Andap, New Bataan Municipality, Compostela Valley Province, a debris flow measuring approximately 1 km wide and 8 km long destroyed an entire village, and 1,080 people were affected: 612 dead and 468 missing. Moreover, a debris flow killed 395 people and left 55 missing, for a total of 450 victims in Baganga City, Dabao Oriental Province. Figure 2.7 [22] shows the central district of New Bataan, Cabinuangan, before the debris flow (Figure 2.7A) and a general view of New Bataan after the debris flow (Figure 2.7B). In Figure 2.7B, red areas are boulder-rich "true debris" flow; orange areas are deposits of more dilute "hyperconcentrated" flows.

2.2 What Happens in Japan

2.2.1 A Disaster-Prone Country

Japan has always been a country prone to natural disasters. The island arc is elongated in the northeast and southwest directions, and the Eurasian and North American plates on which the archipelago is located are constantly being pushed by the Pacific plate from the east and the Philippine Sea plate from the south, resulting in a constant occurrence of earthquakes within the archipelago. The March 2011 earthquake off the Pacific coast of Tohoku was a plate boundary type earthquake with a magnitude of 9.0, and more than 20,000 people were killed or missing [23].

Volcanic activity is also active at plate boundaries. In September 2014, Mount Ontake in Nagano Prefecture erupted, leaving 63 people dead or missing [24].

The Pacific Ocean opens to the east and south of the island arc. Tropical cyclones that develop over the tropical seas of the Pacific Northwest are called typhoons, and since Japan is in the path of these typhoons, it is often struck by them. In 2004, there were 29 typhoons in the Pacific Northwest, and 10 of them made landfall in Japan, the highest number since statistics began in 1951 [25].

At the seasonal transition from spring to summer and again from summer to autumn, the cold Okhotsk Sea air mass in the north and the Ogasawara air mass in the south develop, respectively, and stagnant fronts develop at the boundaries. These fronts are the rainy season front and the autumn rain front. Since the position of the fronts hardly changes, long rains and sluggish weather continue in the same place, and heavy rainfall disasters are likely to occur. When the effects of these fronts are compounded with the effects of typhoons, the damage is even greater. The Kii Peninsula Disaster, the main subject of this book, is a typical example of the damage caused by the combination of an autumn rain front and Severe Tropical Storm Talas (No. 12 in Japan). Since the beginning of history, Japan has been severely affected by these heavy rainfall disasters caused by rainy fronts, autumn rain fronts, and typhoons, and the battle against them is still ongoing.

Figure 2.8 [10] shows the annual average precipitation from 1991 to 2020 as a baseline. Compared to the global average annual precipitation shown in

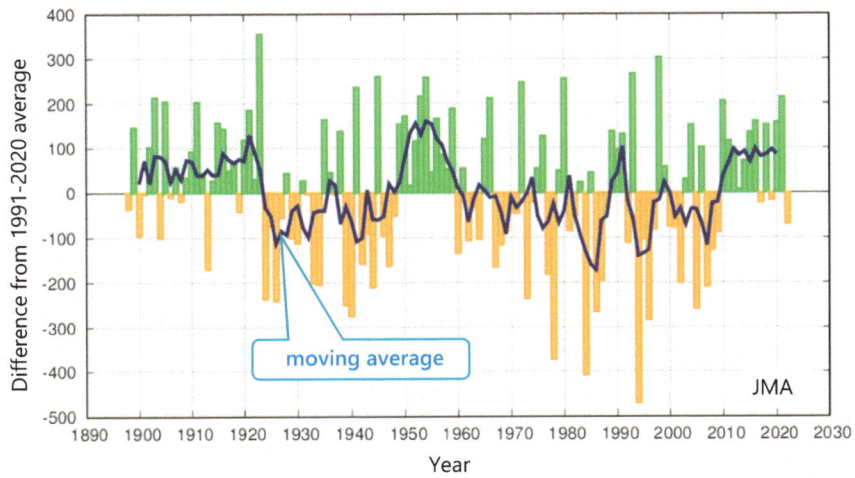

Figure 2.8 Change in average annual precipitation in Japan with respect to the 1991–2020 average [10].

Figure 2.4, it can be seen that dry years and years with high precipitation appear repeatedly. However, if we look at the past decade, it is clear that there have been many high precipitation conditions. These seem to be closely related to the recent trend of frequent sediment and flood disasters. The annual mean temperature variation in Japan was similar to the global trend shown in Figure 2.1.

2.2.2 Sediment Disasters in Japan

As mentioned above, the Japanese archipelago is constantly being pushed horizontally, resulting in few flat areas, and approximately 70% of the country is mountainous. This, combined with the weather conditions, leads to a high incidence of sediment disasters, such as slope failures, debris flows, and landslides (see 7.2.1 for these definitions). Figure 2.9 shows the current trends of sediment disasters in Japan [26]. This figure shows that the average annual number of sediment disasters from 2005 to 2021 was 1,110, indicating that the frequency of sediment disasters has clearly increased over the past 10 years. Although there are year-to-year variations, approximately 80% of the sediment disaster events are slope failures, 15% are debris flows, and 5% are landslides. The trend is that the proportion of debris flows increases as the scale of damage increases.

Although the main focus of this book is the Kii Peninsula Disaster that occurred in September 2011, Table 2.1 summarizes the large-scale sediment and flood disasters that have occurred since then. It can be seen clearly from this table that large-scale sediment and flood disasters that cause many victims occur every year. Major sediment disasters are introduced in the following order. In Table 2.1, notation regarding typhoons, for example, (1112) indicates that this is the 12th typhoon that occurred in 2011

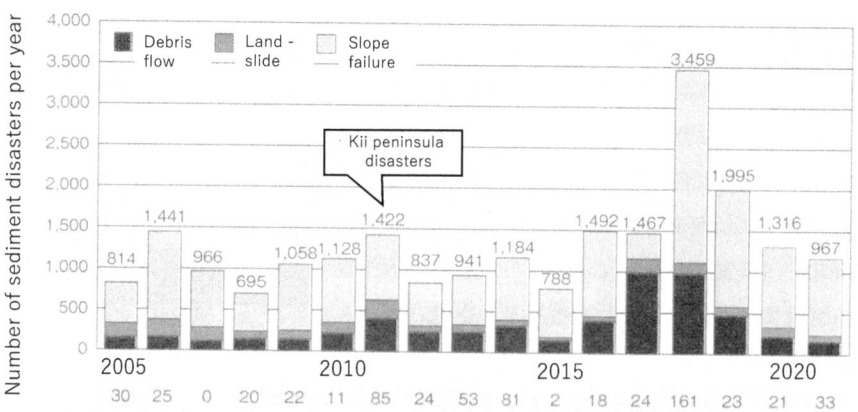

Figure 2.9 Trends of sediment disasters of Japan from 2005 to 2021 [26].

Table 2.1 Recent major sediment disasters in Japan

No.	Year/ Month	Disaster	No. of Victims	No. of Completely and Half-Destroyed Houses	Submerged Houses: Flooding Above and Below Floor Level	Meteorological Conditions	Reference
1	2011/9	Kii Peninsula sediment & flood disasters	98 for total, 88 in Nara, Wakayama, & Mie Pref.	3,538	22,094	Severe Tropical Storm Talas (1112) & autumn rain front	[27,28]
2	2012/7	Sediment disaster in northern Kyushu	33 in Fukuoka, Kumamoto, & Oita Pref.	2,582	10,983	Rainy season front	[29,30]
3	2013/10	Debris flow disaster at Izu-Oshima Island	39	214	-	Typhoon Wipha (1326)	[31,32]
4	2014/8	Debris flow disaster at Hiroshima	77	396	4,164	Localized torrential rain	[33,34]
5	2015/9	Heavy rain disaster at Kanto & Tohoku	14 in Ibaragi, Tochigi, & Miyagi Pref.	7,126	15,654	Extratropical cyclone & Severe Tropical Storm Etau (1518)	[35]
6	2016/8,9	Heavy rain disaster at Tohoku & Hokkaido	29 in Iwate Pref. & Hokkaido	2,903	2,969	Typhoons Lionrock (1610) & Tropical Storm Kompasu (1611)	[36]
7	2017/7	Heavy rain disaster at northern Kyushu	42 in Fukuoka & Oita Pref.	1,422	1,604	Rainy season front and Typhoon Nanmadol (1703)	[37,38]

(Continued)

Table 2.1 (Continued)

No.	Year/Month	Disaster	No. of Victims	No. of Completely and Half-Destroyed Houses	Submerged Houses: Flooding Above and Below Floor Level	Meteorological Conditions	Reference
8	2018/7	Heavy rain disaster in western Japan	231 mainly in Hiroshima, Okayama, & Ehime Pref.	20,663	29,766	Rainy season front & Typhoon Prapiroon (1807)	[39]
9	2019/9	Typhoon disaster in eastern Japan	97 for 13 Pref. including Fukushima, Miyagi, Kanagawa Pref, et al.	14,197	59,038	Extratropical cyclone & Super Typhoon Hagibis (1919)	[40]
10	2020/7	Heavy rain disaster in western Japan	86, mainly in Kumamoto Pref.	6,125	6,971	Rainy season front	[41]
11	2021/7	Mount Izu sediment disaster	28	64	-	Rainy season front	[42]

and affected Japan. The number of fatalities and missing persons and the number of totally or partially destroyed houses and submerged homes in the table are based on information from public organizations. Note that 2022 was a rare year in recent years in which there was not a single sediment or flood disaster that resulted in more than a few casualties. This may be related to the relatively low precipitation in 2022, as shown in Figure 2.8.

The above recent trends in sediment and flood disasters are briefly summarized as follows.

1 the number of victims of sediment and flood disasters has been increasing since 2011.
2 most disasters occur under the influence of typhoons, rainy season fronts, and autumn rain fronts. When these factors combine, the damage is even more extensive (especially 1, 7, 8, 9).
3 linear precipitation zones are also becoming more prominent (especially 2–5, 6–11).
4 vulnerability of volcanic geology, which is also an inherent characteristic of Japan, is also prominent (especially 1–4, 7, 8).
5 Figure 2.10 shows the locations of the sediment and flood disasters listed in Table 2.1. It can be seen that the occurrence of disasters is widely distributed from Kyushu in the south to Hokkaido in the north. In other words, there is a marked tendency for sediment and flood disasters to be widespread. However, we get the impression that the disaster-prone areas are shifting from the south to the north. In the past, damage was most noticeable on the southern coast of the Japanese archipelago, where typhoons regularly hit—such as Kagoshima Prefecture in southern Kyushu, Kochi Prefecture in southern Shikoku, and Wakayama Prefecture in southern Honshu—but

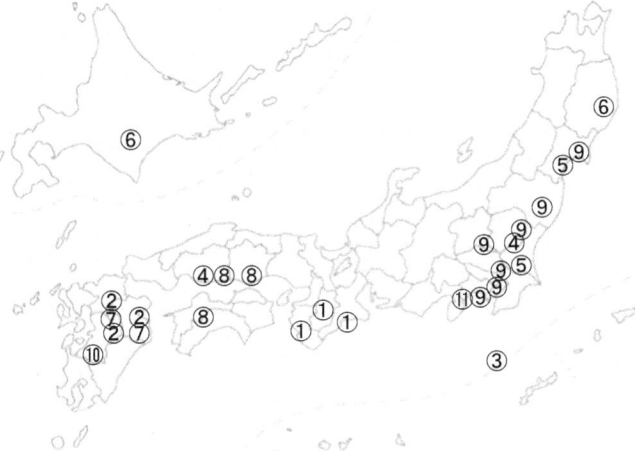

Figure 2.10 Sediment and flood disaster locations shown in Table 2.1.

recently the damage has expanded to the Tohoku and Hokkaido regions. Also in Kyushu, damage was noticeable in Kumamoto in the central part of the island and Fukuoka in the northern part.

This section has introduced recent trends in sediment and flood disasters in Japan. Following is a more detailed description of some of the disasters listed in Table 2.1.

2.2.3 2012 Sediment Disaster at Northern Kyushu [29,30]

From July 11 to 14, heavy rainfall occurred in Fukuoka, Kumamoto, Oita, and Saga prefectures in the northern region of Kyushu. In this region, south of the rainy season front, warm and moist air moved in from the East China Sea, causing extremely unstable atmospheric conditions. In the Kumamoto and the Aso regions of Kumamoto Prefecture, and western Oita, where rain clouds developed in a linear pattern and moved in one after another, heavy rains continued from early morning to late morning on July 12. This phenomenon is known as a linear precipitation zone. In Aso Otohime Area, Aso City, 459.5 mm of rain was observed between 1:00 and 7:00 a.m. on the same day, a record-breaking heavy rainfall. Warm and humid air continued to flow in from the East China Sea from the 13th to the 14th, making the atmospheric conditions extremely unstable. As rain clouds moved in and developed one after another, heavy rainfall occurred mainly in Saga and Fukuoka prefectures on the 13th, and in Fukuoka and Oita Prefectures on the 14th. In Kurogi Area, Yame City, Fukuoka Prefecture, 24-hour precipitation totaled 486.0 mm by 11:30 a.m. on the 14th, and this was the highest record since observations began (1976). Total precipitation for these four days exceeded 500 mm at a total of five stations in the Chikugo Region of Fukuoka Prefecture, the Aso Region of Kumamoto Prefecture, and western Oita Prefecture; and at two stations in the Chikugo Region, it was more than 150% of the monthly average for July.

The heavy rains caused rivers to overflow and debris flows resulting in 33 victims and missing persons in Fukuoka, Kumamoto, and Oita prefectures, as well as 13,565 houses (2,582 damaged and 10,983 flooded) in 4 prefectures including Saga Prefecture. Sediment disasters manifested in 505 cases of slope failure, 256 cases of debris flows, and 76 cases of landslides; the total number was 837. Total of 23 people were killed and 1 person was missing because of sediment disasters. In terms of damage to houses, there were 14 cases of total destruction and 105 cases of half destruction. Photo 2.2 [43] shows the site of a slope failure in the Sakanashi District of Aso City, Kumamoto Prefecture, where six people died.

2.2.4 2013 Debris Flow Disaster at Izu-Oshima Island [31,32]

Typhoon Wipha (1326), which originated near the Mariana Islands at 3:00 a.m. on October 11, moved northward over the South Sea of Japan while developing and approached the coast of the Kanto Region, with a storm area at dawn on

Photo 2.2 Slope failure site in Sakanashi Area, Aso City, Kumamoto Pref., where six
people were killed [43].

October 16 with large size and strong force. The typhoon then moved northward
over the sea east of Kanto Region and changed to an extratropical cyclone off
Sanriku Region at 3:00 p.m. on the 16th. The typhoon and the following extratropical cyclone caused storms and heavy rainfall over a wide area from
western to northern Japan, mainly on the 15th and 16th. Especially in Izu-
Oshima Island, Tokyo, the typhoon brought moist air, which caused rainfall
exceeding 100 mm per hour for several hours from early morning on the 16th,
resulting in heavy rainfall of more than 800 mm in 24 hours. Izu-Oshima Island is
the largest and central island in the Izu Islands, located 120 km south of Tokyo.
    The typhoon caused a large debris flow with driftwoods due to the con-
centration of many slope surface failures in a small area, resulting in extensive

Photo 2.3 Debris flow disasters at Izu-Oshima Island [44].

damage to villages downstream. Total 39 people were killed or missing, and 130 were injured. The number of houses totally or partially destroyed was 214. The surface soil, mainly volcanic ash, which was deposited on top of lava formed by a 14th-century eruption, collapsed. The depth of collapse was generally 1–2 m. Photo 2.3 [44] shows the debris flow disasters at Izu-Oshima Island. Izu-Oshima suffered a major eruption of Mt. Mihara on the same island in November 1986, forcing all islanders to evacuate the island. Therefore, people were highly aware of disaster prevention against volcanic eruptions; however, they were not so aware of debris flows, which led to the spread of damage.

2.2.5 2014 Debris Flow Disaster at Hiroshima [33,34]

From July 31 to August 11, 2014, Tropical Storm Nakri (1412) and Super Typhoon Halon (1411) successively approached the Japanese archipelago. From early August to August 26, a frontal line stagnated near Japan, and warm and very moist air continued to flow into the vicinity of Japan. Therefore, this condition resulted in continuous heavy rainfall-prone weather throughout the country and record-breaking heavy rainfall in many areas. In Hiroshima City, a so-called linear precipitation zone developed, and localized torrential rains continued from the early morning of August 20. Asa-Kita Ward of Hiroshima City experienced a 1-hour maximum rainfall of 121 mm and a 24-hour cumulative maximum of 287 mm. Asa-Minami Ward, Hiroshima City, also experienced the maximum 1-hour rainfall of 87 mm and the maximum 24-hour cumulative total of 247 mm.

The torrential rains caused debris flows in 107 locations and slope surface failures in 59 locations, for a total of 166 locations in Hiroshima City, resulting in 77 fatalities (including 3 related deaths) and 44 injuries. Figure 2.11 [45] is a legible map based on vertical photographs taken on August 28, 30, and 31, 2014, after the Hiroshima debris flow disaster. It shows that many debris flows occurred mainly in Asa-Minami and Asa-Kita Wards. The number of residential houses damaged totaled 4,560, including 396 totally or half collapsed, with extensive damage to railroads and lifelines, especially in the belt-shaped area from Yamamoto Area, Asa-Minami Ward to Obayashi Area, Asa-Kita Ward, where damage was concentrated. The reason the damage in these areas was so extensive is related to the record-breaking precipitation and the geological problems faced by Hiroshima City. The dominant base rock in the affected area is granite, and because granite is easily weathered, the rate of formation of the slope surface layer, which causes surface failure, is rapid. This makes these areas more prone to frequent sediment disasters than areas with base rock that is less prone to weathering. Rainfall patterns were also problematic at this time. There was a period of

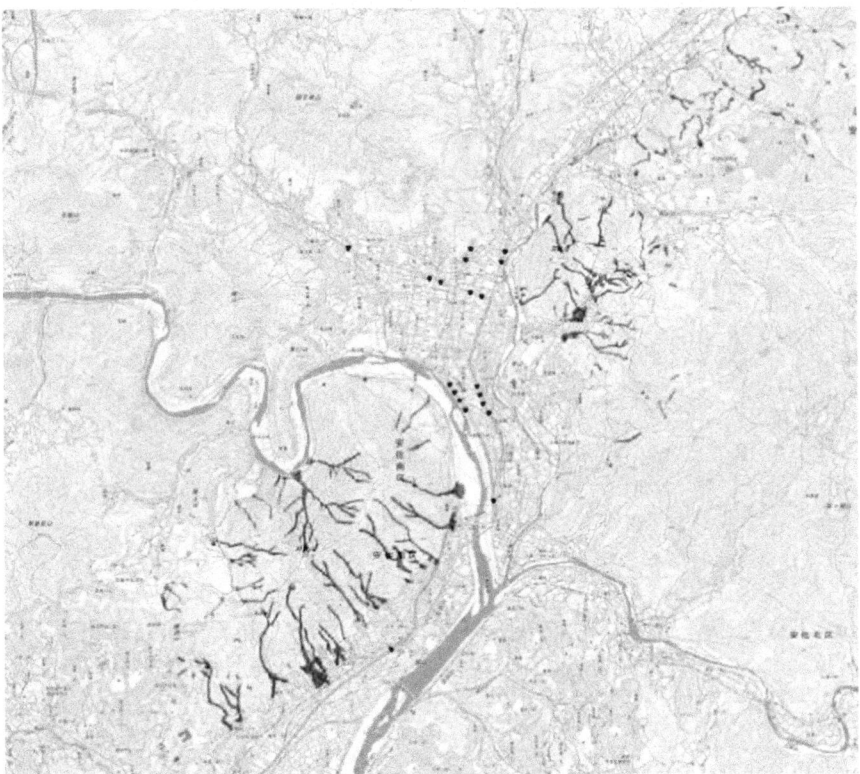

Figure 2.11 Numerous traces of debris flows in Asa-Minami and Asa-Kita Wards, Hiroshima City [45].

decreased precipitation toward midnight on August 19, and many people hesitated to evacuate. Later, heavy rainfall exceeding 100 mm per hour made it impossible to escape even if they wanted to.

2.2.6 2017 Heavy Rain Disaster at Northern Kyushu [37,38]

A rainy season front stalled over western Japan from July 5 to 6, 2017, resulting in heavy rainfall mainly over the northern Kyushu Region. In this region, warm and very humid air flowed toward the rainy front, which stalled near the Tsushima Strait, and a linear precipitation zone formed south of the front. In Fukuoka and Oita prefectures, heavy rain continued to fall from around noon to late at night on the 5th, resulting in record-breaking rainfall. As a result, a special heavy rainfall warning was issued for Fukuoka Prefecture at 5:51 p.m. and for Oita Prefecture at 7:55 p.m. on May 5. The 1-hour precipitation of 129.5 mm was observed in Asakura City, Fukuoka Prefecture, and the 24-hour precipitation of 545.5 mm in Asakura City and 370.0 mm in Hita City, Oita Prefecture, exceeded the normal July precipitation, the highest value since statistics began being kept.

The record-breaking rainfall caused extensive damage in Fukuoka and Oita prefectures, with 42 fatalities and many houses completely or partially destroyed or flooded above floor level. In addition, lifelines such as water and electricity, roads and railroads, and agriculture and forestry, the region's key industries, were severely damaged. In the immediate aftermath of the disaster, more than 2,000 people were forced to live as evacuees. In addition to Fukuoka and Oita prefectures, sediment disasters also occurred in Kumamoto and Nagasaki prefectures, totaling 307 sediment disasters. Sediment disasters occurred as 141 cases of slope failure, 163 cases of debris flows, and 3 cases of landslides. A total of 20 people were killed and 2 were missing because of sediment disasters. Photo 2.4 was taken in Asakura City, Fukuoka Prefecture. One major feature of the 2017 Northern Kyushu torrential rains was the large amount of sediment discharged because of debris flows and slope failures. As seen in Photo 2.4, many riverbeds rose significantly, posing a major disaster-prevention challenge. The worst affected area was Fukuoka Prefecture, and the sediment disaster was characterized by significant damage caused by driftwood.

2.2.7 2018 Heavy Rain Disaster in Western Japan [39]

From June 28 to July 8, 2018, record-breaking heavy rainfall occurred over a wide area nationwide, especially in western Japan. In particular, 76, 124, and 122 observation stations broke the record for the longest rainfall, including 24-hour, 48-hour, and 72-hour precipitation amounts, in that order.

As a result, rivers overflowed and sediment disasters occurred simultaneously over a wide area, mainly in western Japan. The torrential rains caused 223 deaths, 8 people were missing, 20,663 houses were completely or partially destroyed, and 29,766 houses were flooded, resulting in extremely extensive

Photo 2.4 The riverbed, which rose significantly because of the large amount of sediment spilled in the 2017 Northern Kyushu torrential rain disaster, Asakura City, Fukuoka Pref. (Photo by Ryoichi Fukagawa).

damage. According to the Ministry of Land, Infrastructure, Transport and Tourism, there were 1,688 cases of slope failure, 769 cases of debris flow, and 55 cases of landslide. The number of people killed or missing in these sediment disasters totaled 119, the largest number of victims of sediment disasters in Japan in recent years.

One of the most remarkable aspects of the 2018 heavy rain disaster in western Japan was the large number of deaths caused by overflowing rivers. In particular, 51 people were killed in Mabi Town, Kurashiki City, Okayama Prefecture, despite the fact that the flood inundation area was almost identical to the actual inundation area. About 80% of the dead were over 70 years old. How to proceed with evacuation guidance became a serious issue. Photo 2.5 was taken in the Koyaura Area of Sakacho Town, Hiroshima Prefecture, which was hit by a debris flow. In this area alone, 15 people were killed, 1 went missing, and 653 houses were completely or partially destroyed.

2.2.8 2019 Eastern Japan Typhoon Disaster [40]

Super Typhoon Hagibis (1919), which originated on October 6 near Minamitorishima Island, Japan's most easterly island, 1,800 km from Honshu Island, moved westward over the Mariana Islands and temporarily developed

Photo 2.5 Debris flow disasters at Koyaura Area, Sakacho Town, Hiroshima Pref. (Photo by Ryoichi Fukagawa).

into a large and ferocious typhoon, then gradually changed course to the north and moved northward south of Japan, making landfall on the Izu Peninsula with large and strong force before 7:00 p.m. on October 12. It then passed over the Kanto Region and turned into an extratropical cyclone east of Japan at noon on the 13th.

With the approach and passage of Super Typhoon Hagibis, heavy rain, windstorms, high waves, and storm surges occurred over a wide area. Total rainfall from the 10th to the 13th reached 1,000 millimeters in Hakone, Kanagawa Prefecture, and exceeded 500 millimeters at 17 locations, mainly in eastern Japan. In particular, many locations in Shizuoka and Niigata prefectures, the Kanto Koshin Region, and the Tohoku Region saw record-breaking amounts of rainfall, with the highest 3-, 6-, 12-, and 24-hour precipitation amounts ever recorded.

The heavy rains caused flooding in major rivers, such as the Mogami River, as well as many human casualties and much property damage because of sediment disasters and water flowing into low-lying areas. A total of 926 sediment disasters occurred in 13 prefectures, including 294 in Miyagi Prefecture and 138 in Fukushima Prefecture. These included 492 cases of slope failure, 426 cases of debris flow, and 44 cases of landslide. The number of victims of these sediment disasters was 16 dead, 1 missing, and 10 injured,

Photo 2.6 Roads and rivers abandoned by severe debris flows in the 2019 East Japan typhoon disaster, Marumori Town, Miyagi Pref. (Photo by Ryoichi Fukagawa).

while 42 houses were totally destroyed and 58 half destroyed. Photo 2.6 was taken in Marumori Town, Miyagi Prefecture. The paved road and riverbanks were left in pristine condition by the violent debris flow. The debris flow also struck villages downstream, causing extensive human and material damage.

2.2.9 *2020 July Heavy Rain Disaster [41]*

From July 3 to July 31, warm and humid air continuously flowed into Japan due to a front that remained stagnant near Japan, causing heavy rainfall in many areas and resulting in human casualties and property damage. The front was very active, resulting in heavy rainfall in western and eastern Japan, especially in Kyushu, which experienced record-breaking rainfall from July 4 to 7. Heavy rains also fell intermittently around Gifu Prefecture from the 6th, resulting in record-breaking heavy rains from the 7th to the 8th. The Japan Meteorological Agency issued a special heavy rainfall warning for seven prefectures: Kumamoto, Kagoshima, Fukuoka, Saga, Nagasaki, Gifu, and Nagano, urging maximum precautions. The front remained stagnant near Honshu, and rainfall was heavy over a wide area from western Japan to the Tohoku region. Rainfall was especially heavy over the Chugoku Region from the 13th to the 14th and over the Tohoku Region from the 27th to the 28th.

Total precipitation from July 3 to July 31 exceeded 2,000 mm in many places in Nagano and Kochi prefectures, and 24-, 48-, and 72-hour precipitation amounts exceeded the highest values ever observed at many locations in the southern Kyushu, northern Kyushu, Tokai, and Tohoku regions. In addition, the sum of precipitation amounts observed at AMeDAS (Automated Meteorological Data Acquisition System) stations nationwide in early July and the number of occurrences of 1-hour precipitation amounts of 50 mm or more were both the highest since 1982.

Overall, 8 rivers in 7 water systems administered by the national government and 194 rivers in 58 water systems administered by the prefectural government were inundated by bursts or other causes, resulting in extremely extensive damage over a wide area. The damage included 84 deaths, 2 people missing, 25 seriously injured, 6,125 buildings totally or partially destroyed, and 6,971 buildings flooded. Sediment disasters occurred in many areas over a wide area from western Japan to eastern Japan. There were 709 cases of slope failures, 178 cases of debris flows, and 74 cases of landslides, for a total of 961 cases. Kumamoto Prefecture had a particularly high number, with 226 cases, followed by Kagoshima Prefecture with 76, Nagano Prefecture with 73, and so on. The number of victims due to sediment disasters was 16, and 37 houses were totally destroyed and 27 half destroyed. Photo 2.7 was

Photo 2.7 Debris flow disasters at Tagawa Area, Ashikita Town, Kumamoto Pref. (Photo by Ryoichi Fukagawa).

taken at the site of a debris flow in the Tagawa area of Ashikita Town, Kumamoto Prefecture. Three people were killed here.

2.3 Summary

In this chapter, the current status of global warming was presented. The world is clearly experiencing accelerating global warming, and with it, natural disasters such as heavy rains and droughts are becoming more intense, more frequent, and more widespread. Sediment and flood disasters are also inevitably on the rise, which may be explained globally by an increase in the number of such disasters in areas such as Europe and Africa, where these disasters were previously considered relatively rare, and by the emergence of unusual disasters, such as the June 2022 flood disaster in Pakistan. Japan is of course no exception. After the 2011 Kii Peninsula Disaster, which is the main subject of this document, large-scale sediment disasters and flood disasters have continued to occur in many parts of Japan during torrential rains and typhoons, as described in Section 2.2. All of these disasters had a strong impact on us.

References

[1] The Yomiuri Shimbun, 2021. 7. 1. (in Japanese).
[2] The Yomiuri Shimbun, 2021. 7. 17, 18, 19. (in Japanese).
[3] The Yomiuri Shimbun, 2021.8.13. (in Japanese).
[4] The Yomiuri Shimbun, 2021. 9. 4. (in Japanese).
[5] The Yomiuri Shimbun, 2022. 7. 5. (in Japanese).
[6] The Yomiuri Shimbun, 2022. 7. 19. (in Japanese).
[7] The Yomiuri Shimbun, 2022. 9. 1. (in Japanese).
[8] David Wallace-Wells: The uninhabitable earth life after warming, Tim Duggan Books, 2019.
[9] David Attenborough with Jonnie Hughes: A life on our planet, Grand Central Publishing, 2020.10.
[10] JMA: Long-term trends in temperature and precipitation, <https://www.data.jma.go.jp/cpdinfo/temp/index.html> (in Japanese, referenced 2022.9.23)
[11] IPCC WG1 report <https://www.ipcc.ch/report/ar6/wg1/downloads/report/IPCC_AR6_WGI_FullReport.pdf> (referenced 2022.10.20)
[12] International Division, Planning Department, The Sabo and landslide Technical Center: Sediment disasters in the world (No.1 of 2010), Sabo, Vol.103, p.42, 2010. (in Japanese)
[13] EM-DAT <http://www.emdat.be/> (referenced Sep. 23, 2022)
[14] Relief Web <http://www.rliefweb.int/rw/dbc.nsf/doc1007OpenForm> (referenced 2022.9.23)
[15] Reuters AlertNet <http://www.alertnet.org/> (referenced 2022.9.23)
[16] AFPBB Disaster News <http://www.afpbb.com/category/disaster-accidents-crime/disaster> (referenced 2022/9/23)
[17] NHK Online <http://www3.nhk.or.jp/news/?from=tp_an00> (in Japanese, referenced 2022.9.23)

32 *Ryoichi Fukagawa*

[18] Miyase, M. and Kato, N.: Taiwan MORAKOT Typhoon Disaster, Sabo, Vol.101, pp.14–17, 2020.1. (in Japanese).

[19] Kikui, T.: Report on Debris Flow in Zhuguo County, Gansu Province, China, Sabo, Vol.105, pp.16–19, 2011.1. (in Japanese).

[20] AFPBB News: Philippines: Typhoon No. 24 leaves more than 700 dead, 890 missing, <https://www.afpbb. com/articles/-/2916014?pid=9980065> (referenced 2022.9.23)

[21] AMF Lagmay et. al. <https://opinion.inquirer.net/46563/how-debris-buried-compostela-village,Philippine Daily Inquirer/07:49PMFebruary09,2013.> (referenced 2022.9.23)

[22] Kelvin S. Rodolfo et. al.: Super Typhoon Bopha and the Mayo River debris flow disaster, Mindanao, Philippines, 2012.12.<10.5722/intechopen.81669> (referenced 2022.10.13)

[23] Cabinet Office: Disaster Prevention Information <https://www.bousai.go.jp/kohou/kouhoubousai/h23/63/special_01.html> (in Japanese, referenced 2022.10.13)

[24] Cabinet Office: Disaster Prevention Information on Mt. Ontake eruption, <https://www.bousai.go.jp/kohou/kouhoubousai/h26/77/repo_01.html> (in Japanese, referenced 2022.10.13)

[25] Typhoon Information on 2004, https://www.aki-alltech.co.jp/japanese/right9.files/news9_typhoon.htm (in Japanese, referenced 2022.10.13)

[26] Japan Bousaisi Organization: Textbook for disaster prevention specialist, p.55, 2022. (in Japanese)

[27] Cabinet Office: Damage caused by Severe Tropical Storm Talas (1112), 2011.9.28 (in Japanese) <https://www.bousai.go.jp/kaigirep/houkokusho/hukkousesaku/saigaitaiou/output_html_1/pdf/201102.pdf> (in Japanese, referenced 2023.3.3)

[28] Ministry of Land, Infrastructure, Transport and Tourism, Kinki Regional Development Bureau: 2011 Kii Peninsula Great Flood Disaster, January 31, 2014. <https://www-1.kkr.mlit.go.jp/bousai/ qgl8vl0000008 ajd-att/kiihantou-kirokushi.pdf> (in Japanese, referenced 2023.3.3)

[29] Japan Meteorological Agency: Torrential rainfall disaster in northern Kyushu in July 2012, <https://www. data.jma.go.jp/obd/stats/data/bosai/report/2012/20120711/20120711.html> (in Japanese, referenced 2022.9.23)

[30] Ministry of Land, Infrastructure, Transport and Tourism: Occurrence of sediment-related disasters in 2012. <https://www.mlit.go.jp/river/sabo/jirei/h24dosha/h24doshasaigai_gaiyo.pdf> (in Japanese, referenced 2022.9.23)

[31] Japan Meteorological Agency: Wind storms and heavy rain caused by Typhoon Wipha (1326). <https://www.data.jma.go.jp/obd/stats/data/bosai/report/2013/20131014/20131014.html> (in Japanese, referenced 2022.9.23)

[32] Ministry of Land, Infrastructure, Transport and Tourism: Summary of Typhoon Wipha (1326) sediment disaster in 2013. [Preliminary Report] <https://www.mlit.go.jp/river/sabo/h25_typhoon26/typhoon26 sokuhou131021.pdf> (in Japanese, referenced 2022.9.23)

[33] Cabinet Office: Torrential rain disaster since August 19, 2014. https://www.bousai.go.jp/kaigirep/houkokusho/hukkousesaku/saigaitaiou/output_html_1/pdf/201402.pdf> (in Japanese, referenced 2022.9.23)

[34] Erosion Control Division, Public Works Bureau, City of Hiroshima: 8/20 debris flow disaster, Jan. 2015. (in Japanese)

[35] Ministry of Land, Infrastructure, Transport and Tourism: 2015 disaster caused by torrential rains in the Kanto and Tohoku regions. <https://www.bousai.go.jp/

kaigirep/houkokusho/hukkousesaku/saigaitaiou/output_html_1/pdf/201503.pdf>
(in Japanese, referenced 2022.9.23)

[36] Ministry of Land, Infrastructure, Transport and Tourism: 2016 disaster caused by Typhoon in the Tohoku and Hokkaido regions. https://www.bousai.go.jp/ kaigirep/houkokusho/hukkousesaku/saigaitaiou/output_html_1/pdf/201602.pdf> (in Japanese, referenced 2022.9.23)

[37] Cabinet Office: Heavy rain disasters in northern Kyushu. <https://www.bousai. go.jp/kaigirep/houkokusho/hukkousesaku/saigaitaiou/output_html_1/pdf/ 201701.pdf> (in Japanese, referenced 2022.9.23)

[38] Ministry of Land, Infrastructure, Transport and Tourism: Summary of sediment-related disasters at Northern part of Kyushu in 2017 [Preliminary Report No.6]. <https://www.mlit.go.jp/river/sabo/h29_kyushu_ gouu/gaiyou.pdf> (in Japanese, referenced 2022.2.8)

[39] Ministry of Land, Infrastructure, Transport and Tourism: Overview of the July 2018 torrential rain disaster and characteristics of damage. https://www.mlit.go.jp/ river/shinngikai_blog/hazard_risk/dai01kai/dai01kai_siryou2-1.pdf> (in Japanese, referenced 2022.9.23)

[40] Cabinet Office: 2019 Typhoon No.19, 1st Feb. 2020. <https://www.bousai.go.jp/ kaigirep/houkokusho/hukkousesaku/saigaitaiou/output_html_1/pdf/201902_01. pdf> (in Japanese, referenced 2022.9.23)

[41] Cabinet Office: Damage caused by the torrential rainfall in July 2020, 7th Jan. 2021. <https://www.bousai.go.jp/updates/r2_07ooame/pdf/r20703_ooame_40. pdf> (in Japanese, referenced 2022.9.23)

[42] Shizuoka Prefecture: Damage and response to debris flow disaster in the Izusan Area of Atami City [Summary information]. <https://www.pref.shizuoka.jp/_res/ projects/default_project/_page_/001/035/911/atami_sokatsu0624.pdf> (in Japanese, referenced 2022.10.6)

[43] Aso City: Northern Kyushu Torrential Rain Special Edition, Aso City Public Report, 2014.3. (in Japanese)

[44] Oshima Town: 2013 Izu Oshima Sediment Disaster Record, 2017.3. (in Japanese)

[45] Geospatial Information Authority of Japan: Decipherment map based on vertical photographs taken in the 2014 Hiroshima debris flow disaster, 2014.9.4. (in Japanese)

3 Rainfall Characteristics of Severe Tropical Storm Talas and Topographical and Geological Features of the Kii Peninsula

Muneki Mitamura

3.1 Introduction

The record heavy rainfall from Severe Tropical Storm Talas (Typhoon 1112 in Japan) in 2011 caused many disasters in Nara, Wakayama, and Mie prefectures, mainly in the Kii Peninsula in southwest Japan [1,2]. For this area, it was a disaster comparable to the so-called Meiji Totsukawa Torrential Rain Disaster in 1889 [3]. This record rainfall triggered large-scale slope failures (collapse areas of 10,000 m^2 or more, some slope failures accompanied by river channel blockage, so-called deep-seated landslide) and many surface failures on the mountain slopes that form the main part of the Kii Peninsula. In addition, there were many damages caused by debris flows, flooding along rivers, and mountain streams [1,2]. This chapter provides an overview of the rainfall characteristics of Severe Tropical Storm, abbreviated as S.T.S., Talas, which is the main cause of this disaster, and the topographical and geological features of the mountainous areas in the Kii Peninsula as the background of slope disasters.

Although attention tends to focus only on S.T.S. Talas in terms of rainfall characteristics, we think that the preceding rainfall from Typhoon Ma-on (Typhoon 1106) also played a role in the background of the large-scale slope disaster. The major rainfall periods caused by typhoons across the peninsula were July 17–20 for Typhoon Ma-on and August 30 to September 5 for S.T.S. Talas. In the first half of this chapter, rainfall conditions and characteristics of Typhoon Ma-on as well as S.T.S. Talas are presented by organizing meteorological data such as analytical rainfall.

Regarding the geomorphological and geological characteristics, most of the areas where large-scale slope failures occurred are composed of accretionary complexes, which are one of the main constituent geological formations of island arcs. Accretionary complexes are formed in continental marginal areas as a result of oceanic plate subduction and have unique stratigraphic and structural styles. Thus, the fundamentals related to the formation of accretionary complexes should lead to an understanding of mass movements that occur on the major mountain slopes in the island arcs that are composed of accretionary complexes. The latter half of this chapter

DOI: 10.1201/9781003375210-3

reviews the geomorphology of the Kii Peninsula and introduces the structural style of the accretionary complex.

3.2 Rainfall Characteristics Related to the Disaster

3.2.1 Rainfall Caused by Typhoon

Distribution maps of cumulative rainfall and maximum rainfall intensity were created using the analytical rainfall data [4] (Figures 3.1 to 3.4). The distribution of cumulative rainfall for about 50 days between the 2 typhoons (accumulation period from 8 p.m., July 17 to noon September 5) is shown in Figure 3.4. Figure 3.4A shows that the cumulative rainfall is very high in the southeastern mountainous area of the Kii Peninsula. The distribution area of rainfall exceeding 2,500 mm is similar to that of rainfall exceeding 500 mm during Typhoon Ma-on (Figure 3.1) and 1,000 mm during S.T.S. Talas (Figure 3.2), although the paths of the 2 typhoons are different. The cumulative rainfall throughout both periods increased due to similar rainfall distribution zones. Darker inner zones of Figure 3.4A represent the area of rainfall exceeding 3,000 mm. The highest precipitation was in Odaigahara, with values exceeding 4,500 mm. In the mountainous areas, this means that 1.0 to 1.3 times the average annual precipitation for the past 30 years fell in 50 days.

S.T.S. Talas was considered to be the heaviest rainfall on record because the number of AMeDAS stations that set new records were 4 stations for 1-hour rainfall, 17 stations for 24-hour rainfall, and 22 stations for 72-hour rainfall, which is most of the AMeDAS stations shown in Figure 3.5. During Typhoon Ma-on, four 24-hour rainfall records and eight 72-hour rainfall records were broken. The Kii Peninsula Torrential Rain Disaster should be considered in light of the fact that the record was broken twice in 50 days.

Since rainfall intensities did not exceed 100 mm during Typhoon Ma-on, the maximum rainfall intensity distribution over the duration of the 2 typhoons is similar to that during S.T.S. Talas (Figure 3.3). Rainfall intensity of about

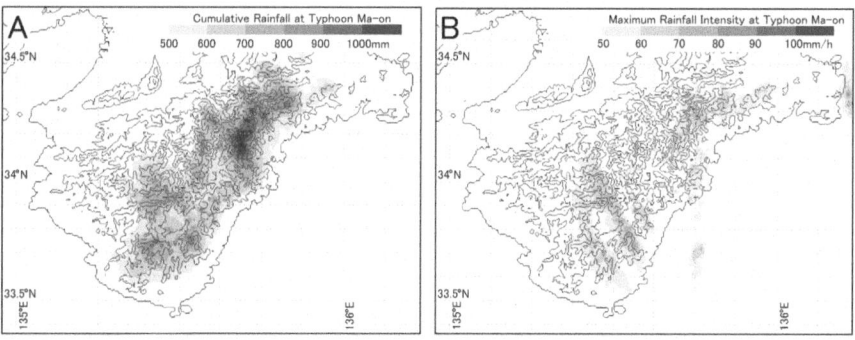

Figure 3.1 Distribution of cumulative rainfall distribution (A) and maximum rainfall intensity (B) during Typhoon Ma-on (Typhoon 1106).

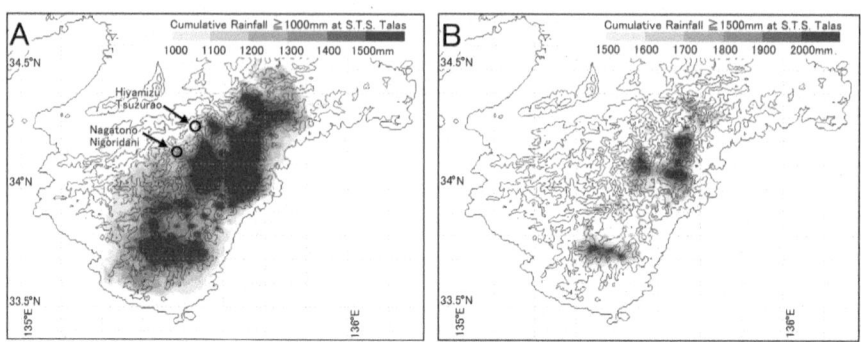

Figure 3.2 Distribution of total rainfall (A: over 1000 mm, B: over 1,500 mm) during Severe Tropical Storm Talas (Typhoon1112).

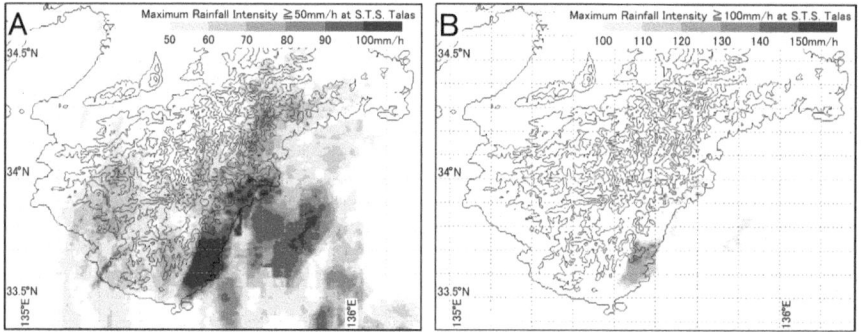

Figure 3.3 Distribution of maximum rainfall intensity (A: over 50 mm/h, B: over 100 mm/h) during Severe Tropical Storm Talas (Typhoon1112).

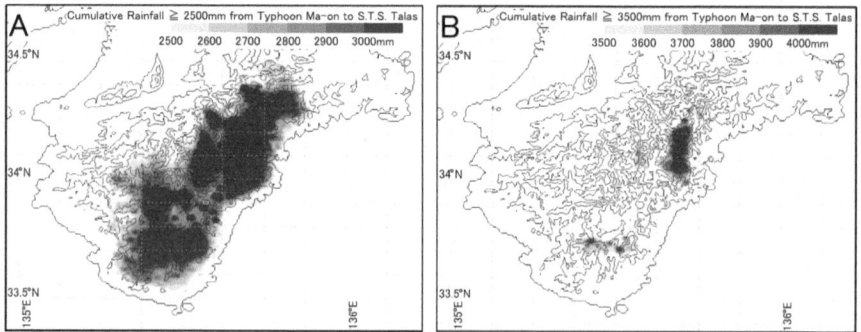

Figure 3.4 Distribution of total rainfall (A: over 2,500 mm, B: over 3,500 mm) from Typhoon Ma-on to S.T.S. Talas.

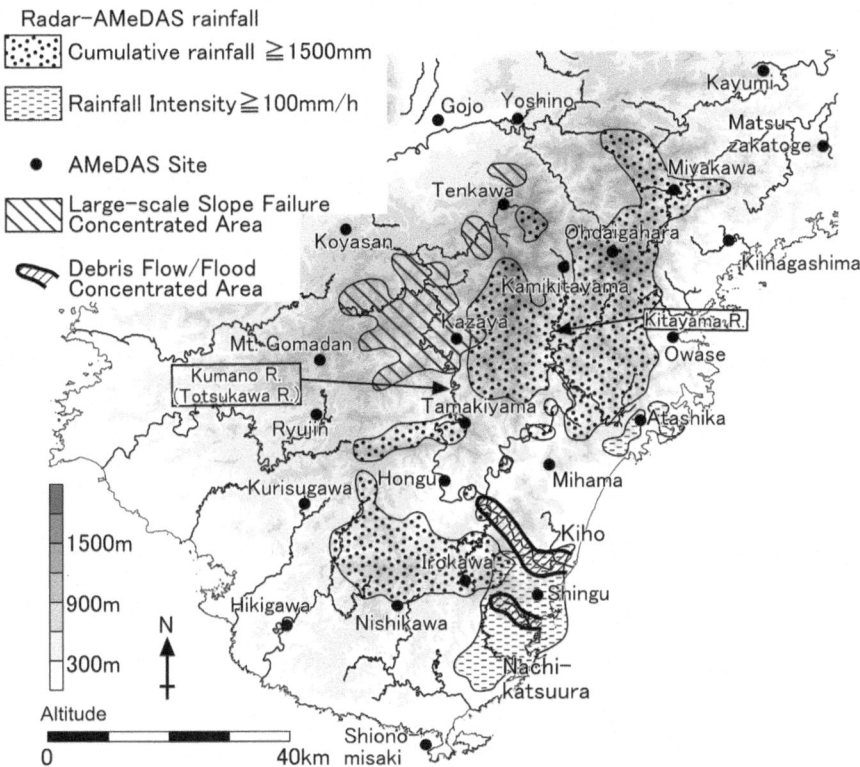

Figure 3.5 Comparison of the distribution area of maximum rainfall and disaster concentration area during S.T.S. Talas.

50 mm/h is widely distributed in both typhoons (Figures 3.1 and 3.3). The distribution of maximum cumulative rainfall and maximum rainfall intensity are different (Figures 3.3 and 3.4). Cumulative rainfall is very heavy in the inland mountainous area, while the maximum rainfall intensity is large in the southeast coastal area of the Kii Peninsula. The distribution of maximum rainfall intensity exceeding 100 mm/h is around Kiho Town, Mie Prefecture, and Shingu City and Nachikatsuura Town, Wakayama Prefecture.

3.2.2 Comparison of Heavy Rainfall Distribution and Disaster Concentration Areas

The concentration area of large-scale slope failure in the southern part of Nara Prefecture falls within the distribution of cumulative rainfall of about 1,000 mm shown in Figure 3.2, but hardly overlaps with the distribution of cumulative rainfall exceeding 1,500 mm (Figure 3.5). It is not simply the amount of rainfall that determines whether or not a slope will fail, but rather the differences in the geological conditions of the slope and its resistance to rainfall.

In contrast, the concentrated areas of debris flow and flood damage that occurred in the lower Kumano River and Nachi River basins generally coincide with the area of maximum rainfall intensity exceeding 100 mm/h. In particular, the area near Nachi Great Falls and Kumano Nachi-Taisha Shrine, both World Heritage sites, where heavy damages occurred, has a maximum rainfall intensity of more than 140 mm/h.

3.2.3 Transition of Rainfall Intensity and Return Periods

In order to examine the characteristics of rainfall during S.T.S. Talas in detail, Odaigahara AMeDAS Station was selected as a representative mountainous area and Shingu Station as a coastal area, and the time series of rainfall recorded at both stations are shown in Figure 3.6. The locations of both stations and the rainfall distribution area are shown in Figure 3.5.

In the mountainous area, as seen in the rainfall time series for Odaigahara, it began to rain lightly around midnight of August 30; 10 to 20 mm/h from the evening of August 31 to midnight of September 1; 30 mm/h from then to midnight of September 2; 40 to 60 mm/h from time to time; then 20 to 30 mm/h until morning of September 4; and stopped around noon of September 4. On the other hand, in the coastal area of Shingu City, it rained on and off until the morning of September 2; rained lightly from then until the morning of September 3; then 10 to 30 mm/h until the morning of September 4, of which

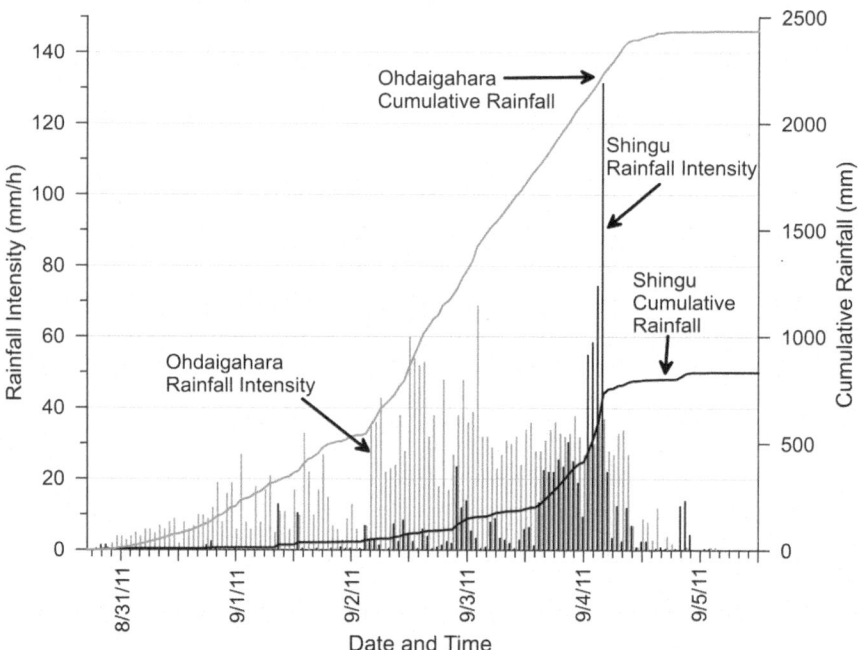

Figure 3.6 Rainfall in mountainous and coastal areas.

50 to 130 mm/h fell heavily from midnight to dawn on September 3; and stopped in the evening of September 4.

Comparing the two stations, the rainfall period of the typhoon was almost the same, six days (actually five days) from the night of August 30 to the evening of September 4, but the rainfall pattern and intensity were different. In Odaigahara, the rainfall intensity was not so high, mainly 20 to 30 mm/h, and it continued for a long time, from the middle to the end of the rainfall period. This resulted in new 24- and 72-hour rainfall observation records. Rainfall exceeding 40 mm/h, which is higher than the rainfall intensity on September 3–4, when many large-scale slope failures took place in mountainous areas, occurred on September 2. In contrast, Shingu experienced very high-intensity rainfall at the end of the rainfall period. This resulted in a new record for rainfall intensity, as well as a new record for 24-hour rainfall and 72-hour rainfall because of the long duration of the S.T.S. Talas.

The recurrence year for each rain duration was determined for representative sites in the mountainous and coastal regions. Probability rainfall intensity curves were generated from the AMeDAS Probability Rainfall Calculation Program [5]. As shown in Figure 3.7, the average recurrence year of the Kamikitayama Village, in the mountainous area, cases with a duration of 12 hours are small, ranging from 5 to 10 years, while those with a duration of over 48 hours exceed 100 years. On the other hand, in the coastal area of Shingu, all the durations far exceeded 50 years and were about 300 years (Figure 3.8).

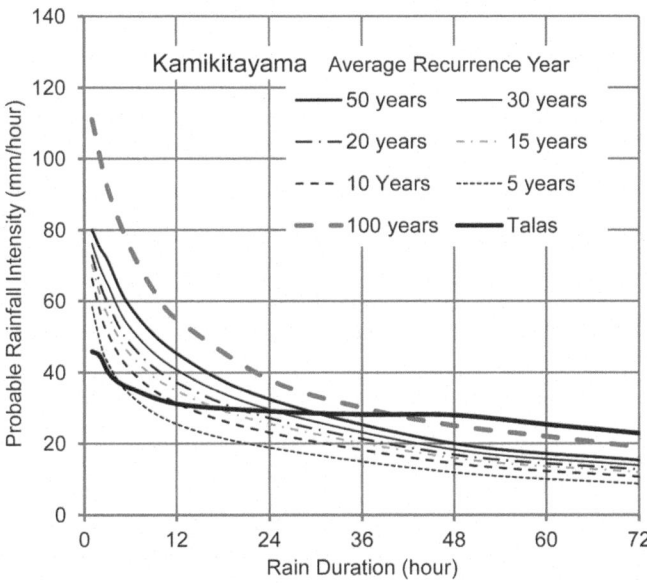

Figure 3.7 Probability rainfall intensity curves at Kamikitayama AMeDAS Station.

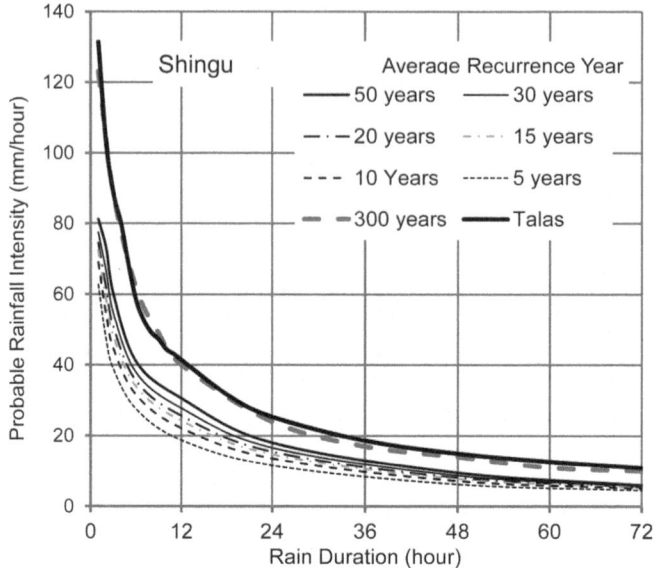

Figure 3.8 Probability rainfall intensity curves at Shingu AMeDAS Station.

Based on these results, it is considered that the large-scale slope failures that occurred in the mountainous areas were not caused by a short period of intense rainfall but by a very long series of rainfalls of 20 to 30 mm/hour. On the other hand, the flooding and debris flow that occurred in Shingu City and Kiho Town in the lower reaches of the Kumano River were probably brought about by the combination of the recent heavy rainfall, which is rare in terms of any rainfall duration, and the preceding runoff from the upstream mountain areas.

3.2.4 *Precursors of Collapse and Preceding Rainfall*

Interviews in Tenkawa Village just after S.T.S. Talas revealed that water had been collected for drinking and other purposes from a spring on the slope in the Hiyamizu site, where a large-scale slope failure occurred, but this water had stopped flowing one month before S.T.S. Talas. This means that at least one month before the large-scale slope failure (around the end of July), deformation had begun in the interior of the slope. Just prior to this, there would be rainfall from Typhoon Ma-on on July 17–20. It is possible that this triggered a change in the slope from a slow degradation progression to a pronounced movement toward destabilization. It became necessary to confirm the rainfall trends of institutions that go back further. Slope failure, channel blockage, and flood damage at Hiyamizu site, Tenkawa Village, are described in Chapter 4.

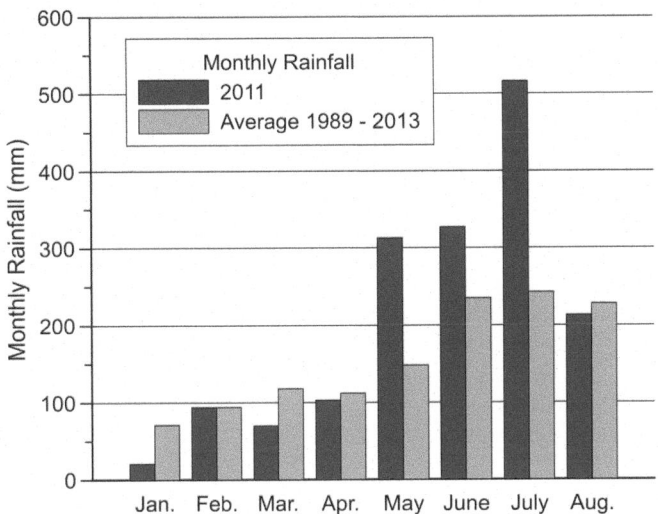

Figure 3.9 Monthly rainfall at Tsudurao AMeDAS Station.

The average monthly rainfall for 25 years and the monthly rainfall in 2011 at Tsuzurao AMeDAS Station near Hiyamizu, (see Figure 4.1 for location) are shown in Figure 3.9. The months of January–April and August are on par with the average, but the months of May–July are 1.4–2.1 times higher than the average (May was affected by Super Typhoon Songda [Typhoon 1102]). This three-month rainfall total was the largest recorded at Tsuzurao Station. This would have been the largest preceding rainfall on record in the same year as Typhoon Talas, which caused the heaviest rainfall on record.

The rainfall from Typhoon Ma-on to S.T.S. Talas based on the analyzed rainfall at Hiyamizu is shown in Figure 3.10. The total rainfall for the 50-day period from July 17, the beginning of Typhoon Ma-on, to September 5, the end of S.T.S. Talas, was 1,890 mm. This corresponds to the average annual precipitation at Tsuzurao Station. It consists of 400 mm during Typhoon Ma-on (duration 48 hours), 420 mm during the period between the 2 typhoons, and 1,070 mm during S.T.S. Talas (duration 120 hours). The rainfall of Typhoon Ma-on is about 40% of that of S.T.S. Talas because of its short duration. The average rainfall intensity relative to duration is 8.3 mm/h for Ma-on and 8.9 mm/h for Talas, which are similar.

The soil rainfall index (three-stage tank model of the Japan Meteorological Agency [6]) was calculated from analytical rainfall at Hiyamizu. The temporal changes from June 17, one month before Typhoon Ma-on including the period of preceding rainfall, to September 5 are shown in Figure 3.11. During S.T.S. Talas, the soil rainfall index was very high with a peak value of 330, 1.5 times the warning value of 220. During Typhoon Ma-on, the index also showed a

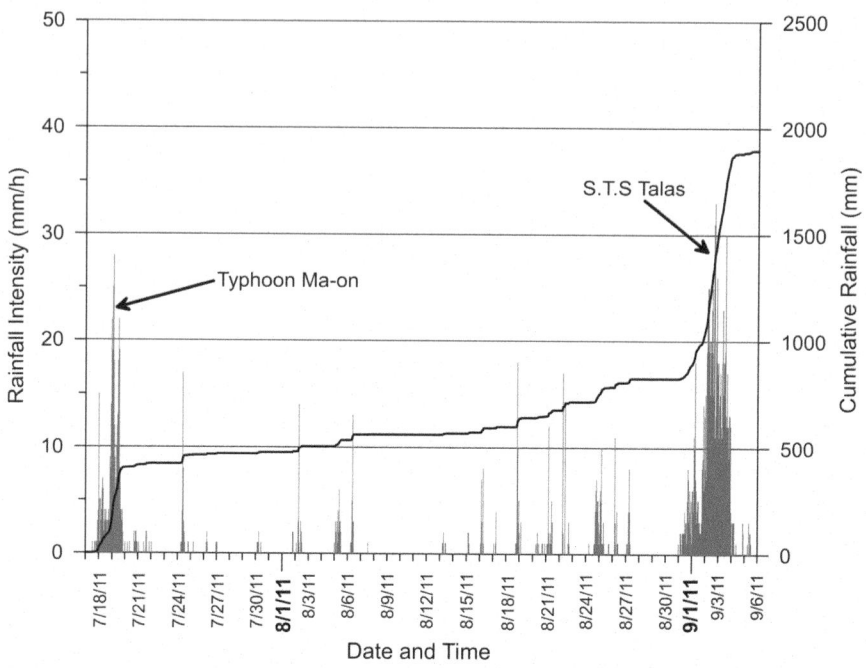

Figure 3.10 Trends of analytical rainfall at Hiyamizu.

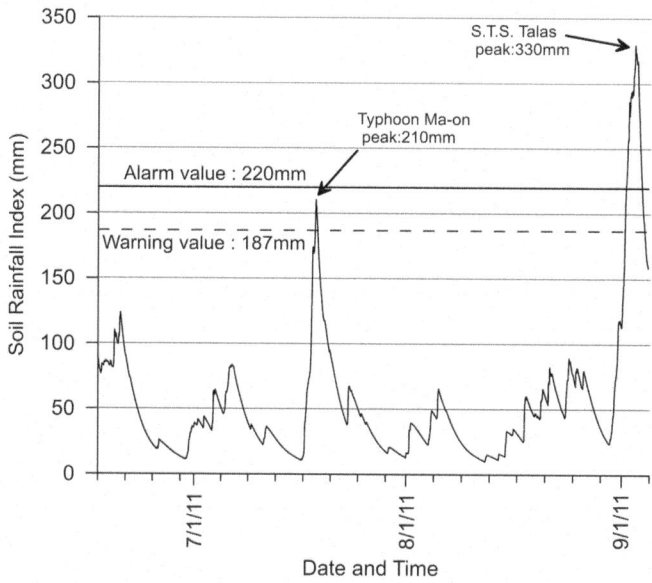

Figure 3.11 Trends of soil rainfall index at Hiyamizu.

high value of 210, almost equal to the warning value. Examination of the soil rainfall index also indicates that there was an impact from preceding rainfall.

3.2.5 Relationship between Rainfall and Large-Scale Slope Failure

In the 54 large-scale slope failures in Nara Prefecture, the approximate dates and times of occurrence are known for 15 of them. Table 3.1 summarizes the rainfall amounts for these 10 large-scale slope failures that occurred in the southern part of the Kumano River Basin. The cumulative rainfall is the total of the analyzed rainfall up to the collapse, and the average annual rainfall is taken from the distribution map of annual rainfall in the Kumano River Basin on the website of the Ministry of Land, Infrastructure, Transport and Tourism [7]. The lower column of the table shows the more southern sites. Annual rainfall tends to be higher in the south. Large-scale slope failures at three sites (Ashinose, Shimizu, and Nojiri) occurred on September 3 during heavy rainfall, while at the other seven sites, slope failures occurred on September 4 after the rainfall subsided. The rainfall during Typhoon Talas exceeded 50% of the average annual rainfall in four to five days, and the cumulative rainfall from Typhoon Ma-on to S.T.S. Talas was almost equal to the annual rainfall.

The cumulative rainfall at each site up to slope failure was similar to or greater than the average annual rainfall (Figures 3.12 and 3.13). At sites with high annual rainfall, the cumulative rainfall leading to slope failure is high. This is thought to be an indication of the slope's resistance to rainfall. The critical line to collapse is about 50% of the average annual rainfall in 5-day total rainfall during S.T.S. Talas. The 50-day rainfall total from Typhoon Ma-on to the collapse is equivalent to the average annual rainfall. Although not shown here, the maximum rainfall intensity is similar or slightly higher

Table 3.1 Rainfall Amounts for 10 Large-Scale Slope Failures

Large-Scale Slope Failure Site	Cumulative Rainfall of S.T.S. Talas Until Collapse (mm)	Cumulative Rainfall from Typhoon Ma-on to S.T.S. Talas (mm)	Average Annual Precipitation (mm)
Ashinose	880	1,610	1,650
Tsubonouchi	1,040	1,850	1,650
Hiyamizu	1,060	1,880	1,650
Tsujido	1,030	1,800	1,730
Ui (Shimizu)	1,160	2,020	1,750
Hinoseyama	1,080	2,010	1,800
Kitamata	980	1,860	1,900
Nigoridani	1,200	2,080	1,950
Ohkuzure	1,140	2,030	1,950
Nojiri	1,250	2,340	2,400

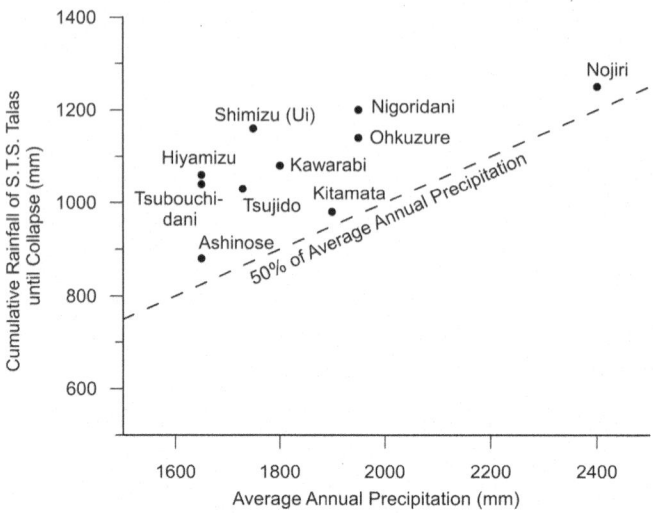

Figure 3.12 Relationship between cumulative rainfall and average annual rainfall during S.T.S. Talas.

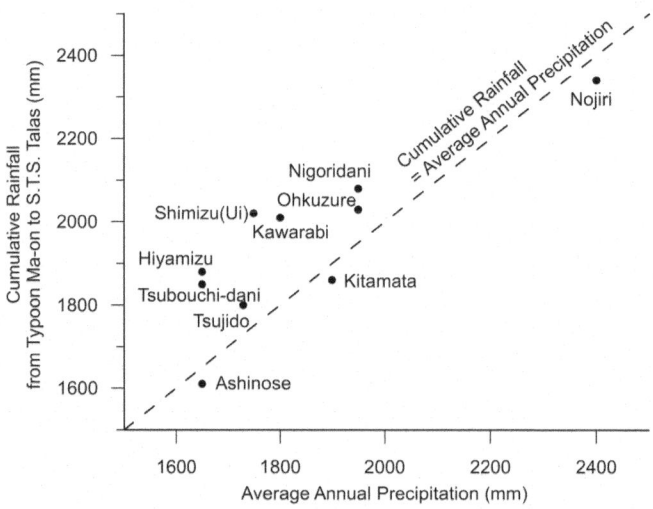

Figure 3.13 Relationship between cumulative rainfall and average annual rainfall from Typhoon Ma-on to S.T.S. Talas.

for Typhoon Ma-on than for S.T.S. Talas. Thus, maximum rainfall intensity is not a direct factor in large-scale slope failures in the sites discussed here.

It seems that rainfall and geological conditions are not the only factors that lead to large-scale slope failures. Seven of the ten sites collapsed when the rainfall had calmed down. For example, the slope failure at Hiyamizu occurred 5.5 hours after the strong rainfall of more than 10 mm/h had stopped. When the rainfall calmed down, it seems that seepage water and pore water inside the slope somehow acted significantly on the unstable side, leading to the collapse. In other words, the collapse could have occurred even if the rain had stopped a little earlier, instead of being a long continuous rainfall. In this case, the limit line indicated earlier would not be the threshold.

Figure 3.14 shows the change over time of the soil rainfall index at the collapse at the Nigoridani site (see Figure 4.1 for location). This site is similar to the Hiyamizu site as described above, but shows higher values during Typhoon Ma-on and S.T.S. Talas due to higher rainfall. The soil rainfall index at this site is similar to that of the Hiyamizu site and is higher during Typhoon Ma-on and S.T.S. Talas because of the higher rainfall. The collapse occurred 8 hours after the peak of the soil rainfall index during S.T.S. Talas. The alarm value was exceeded late at night on September 1, and the collapse occurred before dawn on September 4. There was a 53-hour grace period between the two dates, and if the evacuees had been prepared since September 2 and had been able to evacuate to a safe place until the morning on September 3, their lives would have been spared. In cases such as the present one, where rainfall of 20 to 30 mm/h is continuous for a long period of time, it is considered effective to watch the soil rainfall index and time the evacuation according to when the warning or alarm value is exceeded.

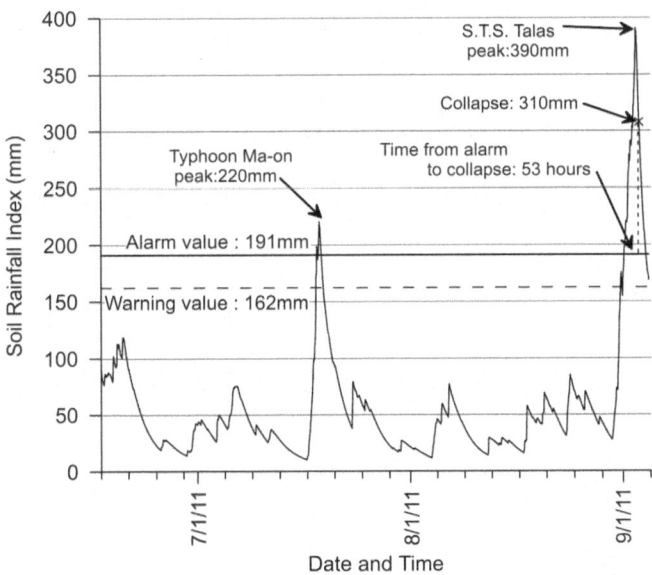

Figure 3.14 Trends in soil rainfall index at Nigoridani site.

3.3 Topography and Geology

3.3.1 Topographical Outline

The Kii Peninsula is bordered on the north by the Kino River (Yoshino River), which originates in the Daiko Mountains and flows west to the Wakayama Prefecture side, and the Kushida River, which flows east to the Mie Prefecture side (Figure 3.15). The Kii Mountains, whose highest peak is Mount Hachikyogadake (1,915 m above sea level), are located in the central part of the Kii Peninsula and rise to over 1,000 m above sea level. The Omine Mountains, which extend from north to south in the center of the Kii Mountains, are a series of mountain ridges over 1,700 m above sea level. To the east of the Ohmine Mountains, the Daiko Mountains extend from north to south. The Koya-Gomadan Mountains, which are more than 1,500 m above sea level, are distributed on the west side of the Ohmine Mountains. Thus, the main ridges of the main part of the Kii Mountains show a north-south extension. Most of the ridges derived from the main ridge show an east-west extension and developed under the influence of the main geological

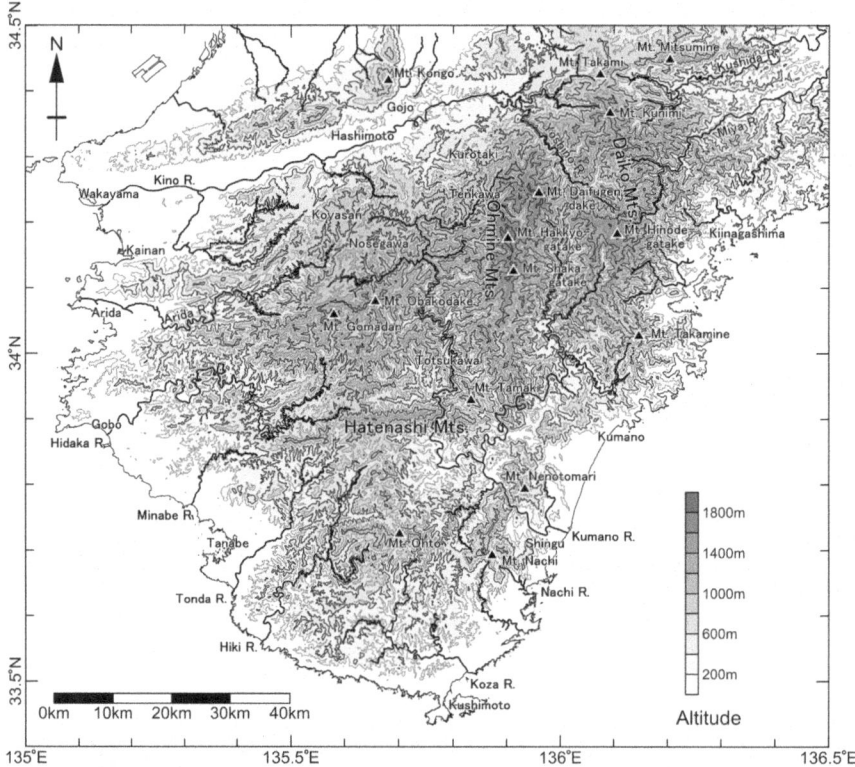

Figure 3.15 Topographic map of the main part of the Kii Peninsula (created from SRTM-3 data [8]).

structure of the Shimanto accretionary complex distributed in the Kii Peninsula (Figure 3.15).

For this reason, major river systems flowing through the Kii Peninsula, such as the Kino River (Yoshino River), Arita River, Hidaka River, Kushida River, and Miya River, show east-west flow directions. The Kumano River system, the largest river system in the Kii Peninsula, divides into the Totsu River and the Kitayama River in its middle reaches and flows between the Ohmine and Daiko Mountains and the Koya-Gomadan Mountains. Therefore, in the middle and upper reaches of the Kumano River, the specific height from the main ridge to the bottom of the river valley is over 1,000 m, and steep mountain slopes are formed on both sides of the river valley. The main channel of the Kumano River takes a north-south flow path dividing the mountainous terrain, while the tributary rivers flowing into the main channel are influenced by the major geological structures of the accretionary complex, forming an east-west flow system.

The southern part of the Kii Peninsula, south of the line connecting Gobo and Kumano, has a mountain massif with a slightly lower elevation (of about 1,000 m) than the central part of the Kii Peninsula. The southern part of the Kii Peninsula south of the line connecting Gobo and Kumano forms a slightly lower mountain massif (up to about 1,000 m in elevation) than the central part of the Kii Peninsula. Mount Takamine and the surrounding mountains extend north of Kumano. Low undulating mountains up to about 200 m in elevation extend along the coast of the Kii Peninsula. Coastal terraces of about 20 m in elevation, which were formed after the Late Pleistocene, have developed along the coast. Alluvial lowlands extend along the estuaries of major rivers. The rest of the coast forms a rias-type rocky coast. The Quaternary formation is distributed on lowlands along the coast and consists of unconsolidated layers of gravel, sand, and mud, forming terrace and alluvial surfaces.

3.3.2 Geological Outline

The Kii Mountains consist mainly of accretionary complexes formed during the Jurassic to Paleogene periods [9–11]. The main geological structure shows an east-west extension. In contrast to the Inner zone on the northern side of the Median Tectonic Line, which runs along the Kino River from east to west in the north of the Kii Peninsula, the southern part of the line is called the Outer zone. The Outer zone is structurally divided into the Sambagawa, Chichibu, and Shimanto Terranes, in that order from north to south, based on the formation ages and constituent geology of the accretionary complexes that comprise each terrane (Figure 3.16).

In the southern Kii Peninsula, sedimentary rocks of the Tanabe and Kumano Groups of Miocene age, which were deposited in basins formed in the coastal to semi-pelagic regions, unconformably overlie the above accretionary sedimentary rocks. Kumano acidic igneous rocks erupted by subsequent volcanic

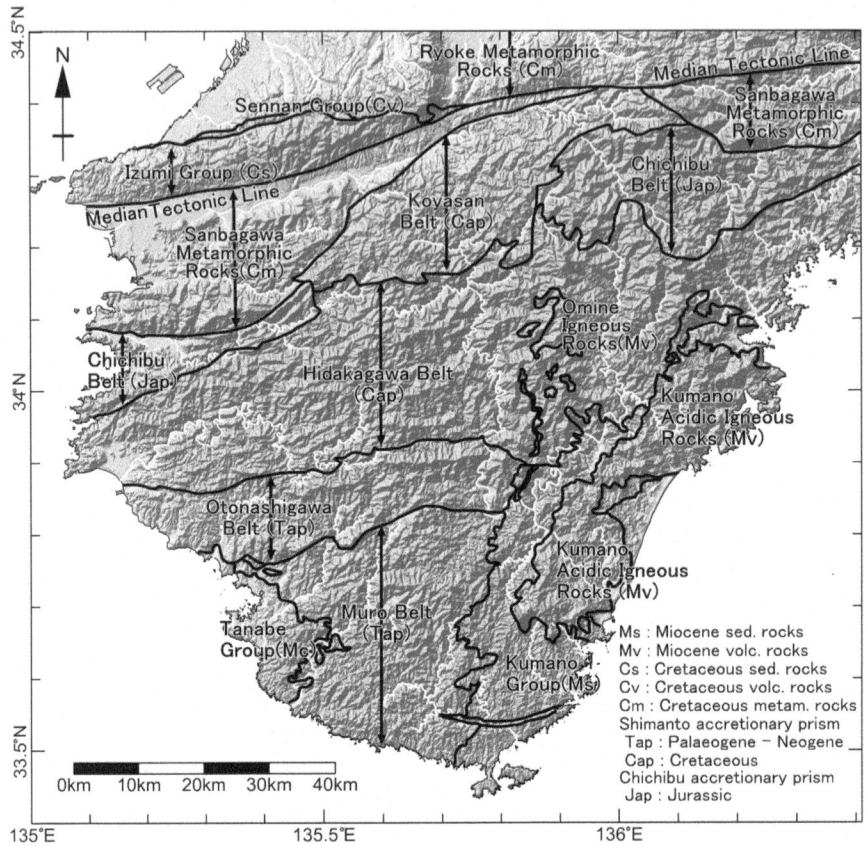

Figure 3.16 Topographic relief and geologic schematic map of the main part of the Kii Peninsula (created from SRTM-3 data [8] and Seamless Geological Map [12]).

activity are distributed through the Kumano Group. In the Quaternary, under the influence of the uplift of the Kii Peninsula and sea-level changes, unconsolidated sedimentary layers were deposited along the coastal areas forming flat surfaces, and alluvial plains and coastal terraces were formed along rivers.

3.3.3 Structure of Accretionary Complex

The large-scale slope failures in the Kii Mountains caused by the 2011 Kii Peninsula torrential rainfall mainly occurred on slopes composed of accretionary complexes. This section presents an overview of the formation of accretionary complexes and their characteristics.

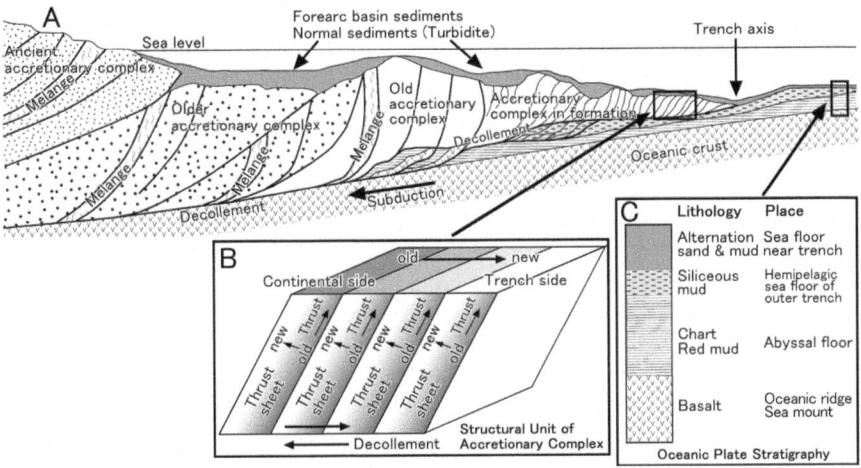

Figure 3.17 Structural overview of the accretionary complex (adapted from Koide, 2013 [13]).

Accretionary complex (accretionary prism) is a geologic body formed by the subduction of an oceanic plate beneath a continental plate, with sediments and other materials on the oceanic plate added to the margins of the continental plate. An overview of the accretionary complex is shown in Figure 3.17 [13].

Accretion occurs in the region landward of the trench where oceanic plate subduction begins (left side of the trench axis in Figure 3.17A). At the Central Ridge, oceanic crust is created here through basaltic volcanism fed by the mantle. The production of oceanic plates, their separation from the ridge, cooling and solidification, and the increase in plate thickness result in the formation of submarine topography that gradually deepens from the ridge. Since there is little supply of terrestrial source material on the seafloor in these pelagic regions, the remains of planktonic and benthic microorganisms are deposited. At depths greater than 4,000 m, the remains of calcareous-shelled microorganisms dissolve in seawater due to water pressure. In such an environment, siliceous microbial remains are mainly deposited and chert layers are formed. In hotspots such as the Hawaiian Islands (where basaltic volcanism occurs in spots), volcanic islands are formed in the pelagic zone. Limestone is deposited in the shallow-water areas around volcanic islands, and chert is deposited on their deep-water slopes. The oceanic plate approaches the continent, carrying pelagic sediments and basalt derived from volcanic islands. In the near-land regions of the oceanic plate, mud and sand supplied from land accumulate over pelagic sediments. In the vicinity of the trench, the oceanic plate is overlain by basaltic rocks, chart and limestone, mudstone, and sandstone (sandstone-mudstone alternation) in ascending order. Such a sequence is called oceanic plate stratigraphy (Figure 3.17C).

As the oceanic plate subduction, the sediments on the oceanic plate are stripped away to form accretionary complexes. The pelagic, semi-pelagic, and trench sediments deposited on the oceanic plate during subduction are stripped away, creating reverse faults on the margins of the continental plate (Figure 3.17B). During accretion, sediments on the oceanic plate are structurally disturbed, creating a rock complex of mixed blocks of basaltic rocks, chart, and limestone in muddy matrix. Such rock complexes with a significant degree of shear and disturbance are called mélange.

The continental plate margins are accreted by sediments and volcanic rocks stripped from the oceanic plates via numerous reverse faults. Basically, the geologic structure is formed with the imbricate structure tilted toward the landward side. In the individual sheet-like geologic bodies (thrust sheets) bounded by reverse faults, the landward side is basically higher (younger). On the other hand, a comparison of adjacent thrust sheets shows that the more landward from the trench, the older the blocks. The thrust sheet on the trench side is located lower and has a relatively younger formation age (Figure 3.17B).

The fault plane at the bottom of the reverse fault becomes nearly horizontal, and the reverse faults are connected to each other to form one large fault that forms the base of the thrust zone. This fault, called the decollement, is the boundary between the continental and oceanic plates. With the development of reverse faults, the accretionary body is compressed, lateral shortening occurs, and the old thrust sheet is lifted and tilted significantly. This process creates new, larger-scale reverse faults. Such reverse faults disturb the basic structure when accreted in the trench and are called out-of-order thrusts. Out-of-order thrusts not only cause large-scale rupture of accretionary complexes, but also cause earthquakes and the formation of sedimentary basins (forearc basins) along island arcs. Semi-bathyal to shallow-marine sediments (normal sediments) are deposited in forearc basins (Figure 3.17A).

Accretionary studies, along with the development of plate tectonics theory, have focused on understanding the Mesozoic and Paleozoic sedimentary complex of the Japanese Islands. The geologic structure of the accretionary complex is becoming better understood as a result of surface geological investigations and the results of large-scale seismic surveys and deep-sea drilling. Currently, the Nankai Trough earthquakes are occurring repeatedly due to the subduction of the Philippine Sea Plate under the Eurasian Plate. The subduction of oceanic plates at continental margins has formed accretionary complexes while generating plate boundary earthquakes, and steep mountain ranges composed of accretionary complexes have been formed with deformation and uplift. In these mountainous areas, the pronounced sheared rock complexes are a factor in causing large-scale slope failures.

3.3.4 Summary

Severe Tropical Storm Talas (August 30 to September 5) in 2011 brought record rainfall totaling more than 1,000 mm over a wide area of the Kii Peninsula. Large-scale slope failures in and around the Kii Mountains and river flooding and debris flows in the southeastern coastal areas led to large-scale disasters across the Kii Peninsula. The first half of this chapter presents rainfall characteristics related to heavy rainfall disasters in the Kii Mountains. The second half outlines the topography and geology of the Kii Peninsula and introduces the structural patterns of accretionary complexes that mainly compose the Kii Mountains.

In the Kii Mountains, S.T.S. Talas brought rainfall intensity of 20 to 30 mm/h for a long period of time from the middle to the end of the period, causing large-scale slope failures. On the other hand, in the southeast coastal lowlands, rainfall exceeding 40 mm/h at the end of the rainfall period caused river flooding and debris flow damage. Along with the rainfall from Typhoon Talas, the rainfall and soil rainfall index indicate that the preceding rainfall from Typhoon Ma-on (July 17–20) was one of the factors destabilizing slopes. Based on the trends in the soil rainfall index, in the case of continuous rainfall of 20 to 30 mm/h, it is considered effective to measure evacuation, and when the warning or alarm value of this index is exceeded. The Kii Mountains are a mountain range with elevations exceeding 1,500 m above sea level. The main geological structure of the accretionary complex, represented by the Shimanto Terrane, is influenced by the east-west-oriented ridges that are derived from the main ridges. The Kii Mountains have undergone significant deformation and uplift from the repeated growth of accretionary zones since the Jurassic period. Significantly sheared composite rock units are the cause of large-scale slope failures.

References

[1] Geotechnical Engineering Society of Japan, Kansai Branch: Investigation Report of Heavy Rain Disaster in Kii Peninsula by Typhoon 1112 (Talas) in 2011, p.201, 2011. (in Japanese)

[2] Geotechnical Engineering Society of Japan Kansai Branch, Kansai Branch of the Japan Society of Applied Geology, Kansai Geological Surveyors Association, and Chubu Geological Surveyors Association: Report of the Research Committee on Response to Ground Disasters Caused by "Unexpected" Heavy Rainfall, p.404, 2015. (in Japanese)

[3] Hirano, M., Suwa, H., Ishii, T., Fujita, T, and Gocho Y.: *Reexamination of the Totsukawa Hazard in August 1889 with Special Reference to Geologic Control of Large-scale Landslides*, Annuals of disaster prevention Research Institute Kyoto University, No.27, B-1, pp.369–386, 1984. (in Japanese)

[4] Japan Meteorological Agency: Analytical Rainfall Record, in 2011, Japan Meteorological Service Support Center, DVD media, 2012. (in Japanese)

[5] Hydraulic Engineering Research Group, Public Works Research Institute: AMeDAS Probability Rainfall Calculation Program. <http://www.pwri.go.jp/jpn/results/offer/amedas/top.htm>, (in Japanese, referenced 2013.3.1)

[6] Japan Meteorological Agency: soil rainfall index. <http://www.jma.go.jp/jma/kishou/ know/bosai/dojoshisu.html>. (in Japanese, referenced 2014.2.15)

[7] River Bureau, Ministry of Land, Infrastructure, Transport and Tourism: Overview of the Shingu River Basin and River System, pp.2–4, 2008. http://www.mlit.go.jp/river/shinngikai_blog/shaseishin/kasenbunkakai/shouiinkai/kihonhoushin/080221/pdf/refl-3.pdf. (in Japanese, referenced 2013.3.1)

[8] Jet Propulsion Laboratory, NASA: Shuttle Radar Topography Mission (SRTM). <http://www2.jpl.nasa.gov/srtm/> (referenced 2014.9.4).

[9] The Geological Society of Japan: *Regional Geology in Japan Kinki District-*, Asakura Publishing Co., Ltd., p.448, 1999. (in Japanese)

[10] Yoshimatsu, T., Nakaya, S., Kodama, T., Terai, K. and Harada, T.: Geology and hot springs in the Kii Peninsula, Urban Kubota, Vol. 38, p.35, 1999. (in Japanese)

[11] Kishu Shimanto Research Group: New Developments in the Study of the Shimanto Accretionary Complex in the Kii Peninsula, *Monograph of the Association for the Geological Collaboration in Japan*, No. 59, p.295, 2012. (in Japanese)

[12] National Institute of Advanced Industrial Science and Technology: Seamless Geological Map. <https://gbank.gsj.jp/seamless/> (in Japanese, referenced 2015.1.14).

[13] Koide, Y.: *Formation and Disturbance Mechanism of Accretionary Wedge in Island Arcs, Journal of the Society of Humanities*, Sapporo Gakuin University, No. 93, pp.37–58, 2013. (in Japanese)

4 Disasters in Nara Prefecture

Muneki Mitamura

4.1 Introduction

Total precipitation in the Kii Peninsula associated with Severe Tropical Storm Talas (Typhoon 1112) from 5 p.m. on August 30 exceeded 1,000 mm over a wide area, with 72-hour rainfall reaching 1,652 mm and total rainfall 1,808 mm at an AMeDAS station located in Kamikitayama Village, Nara Prefecture. The analytical rainfall exceeded 2,000 mm near Ohdaigahara [1,2].

The human casualties of this disaster in Nara Prefecture are summarized as follows: 14 people were killed, 10 people are missing, and 359 households were evacuated at the peak of the disaster [1,2]. In particular, about 3,000 slope failures occurred on the mountain slopes of the Kii Peninsula. In the southern area of Nara Prefecture, there were more than 1,800 collapses, and 54 large-scale slope failures (with an area of more than 10,000 m^2 and some slope failures with channel blockage, so-called deep-seated landslide) were observed. Total 45 sites occurred in the Shimanto Terrane of Cretaceous to Paleogene accretionary complexes [1].

Large-scale slope failures, Akadani, Nagatono, Ui (Shimizu), Tubonouchi, and Nojiri, occurred on mountain slopes in the Kumano River Basin (Figure 4.1) [1]. These collapses resulted in major disasters from debris flows, channel blockages (16 locations), and rising river levels and bore propagations. This chapter describes several large-scale slope failures in the Nara Prefecture, examples of surface failures and their mechanisms, and typical locations of river damage, based on the survey and research report [3].

4.2 Large-Scale Slope Failure

In the Kumano River Basin in the southern area of Nara Prefecture, many large-scale slope failures have occurred in the Cretaceous accretionary complex of the Hidakagawa Terrane (Northern Shimanto Terrane) [5,6]. Large-scale failures are concentrated in the Miyama Formation [5] (Miyama Accretionary Complex [6]) and Hanazono Formation [5] (Hanazono Accretionary Complex [6]). The Miyama Formation is a fractured unit, and the muddy complex part with blocks of chert and basalt is particularly fractured. The Hanazono

DOI: 10.1201/9781003375210-4

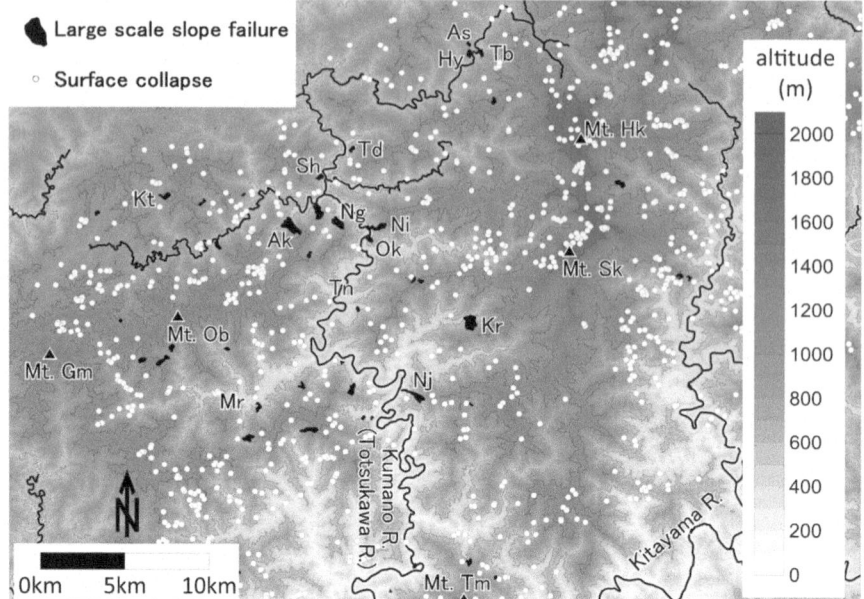

Ak: Akadani, As: Ashinose, Hy: Hiyamizu, Kr: Kuridaira, Kt: Kitamata, Kw: Kawarabi,
Mr: Miura, Ni: Nigoridani, Nj: Nojiri, Ng: Nagatonodani, Ok: Ohkuzure, Sh: Shimizu(Ui),
Tb: Tsubonouchi–dani, Td: Tsujido, Tk: Tenkawa, Tn: Tanise, Mt.Gm: Mt. Gomadan,
Mt. Hk: Mt. Hakkyogatake, Mt. Ob: Mt. Obakodake, Mt. Sk: Mt. Shakagatake,
Mt. Tm: Mt. Takakiyama

Figure 4.1 Distribution of slope failures in the southern region of Nara Prefecture.
The location of the slope failure area is based on GIS point data of the
slope failure site by the Nara Prefecture Large-Scale Landslide Disaster
Survey on the 2011 Kii Peninsula Flood [4].

Formation is also the accretionary complex that formed at almost same time as
the Miyama Formation. The Hanazono Formation, which reaches deeper into
the crust, is recognized to have been placed north of the Miyama Formation
by the Yanase thrust [5]. The Hanazono Formation has suffered significant
shear deformation, and the shale has developed particularly pronounced
cleavage and shear planes. The followings are the geological conditions of the
large-scale slope failure areas where field surveys were conducted.

4.2.1 Tsubonouchi Area in Tenkawa Village

The three large-scale slope failures in Tubonouchi Area, such as Ashinose,
Tubonouchi Valley, and Hiyamizu, occurred on mountain slopes along the
Tennokawa River (Kumano River) (Photo 4.1, Figure 4.2). According to the
Landslide Geomorphology Database of the National Research Institute for
Earth Science and Disaster Prevention, Japan [7], these slopes had the char-
acteristics of slope variation landforms. The shape of the slopes was gentle in the

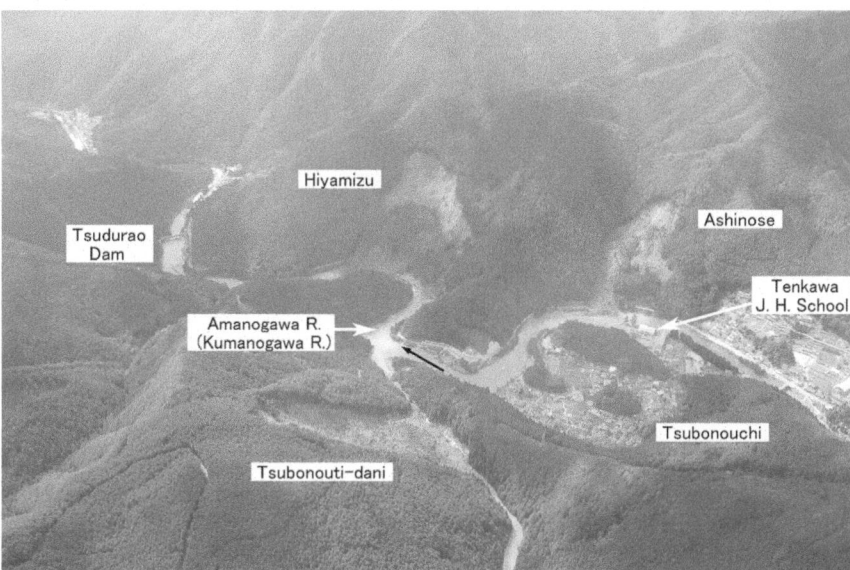

Photo 4.1 Panoramic view of the damage in the Tsubonouchi area of Tenkawa Village (Photo provided by Tenkawa Village).

middle part, and the lower part formed steep slopes of 40 to 50 degrees toward the riverbed. The slope failure of Ashinose occurred at about 8:30 p.m. on September 3. The collapsed sediment moved southeastward and reached the southern slope of the valley. The sediments were deposited in the valley and flowed down to the Tennokawa River as a debris flow. The slope failure of Tsubonouchi Valley occurred at about 8:30 a.m. on September 4 [1,2]. The sediments from this collapse reached the other side of the valley and blocked the river channel. The slope failure of Hiyamizu occurred after 12:30 p.m. on September 4 [1,2]. The sediment by this collapse flowed directly into the Tennokawa River, causing the river channel to become blocked. The impact of these collapsed sediments on rivers will be shown in the later section on river disasters.

The Hanazono Formation, which mainly comprises the mountains of the Tsubonouchi Area, consists of basalt, basaltic tuff with red shale (pelagic submarine volcanic and marine sediments), shale-dominated, and sandstone-shale alternation layers, in ascending order. These strata constitute the oceanic plate stratigraphy (Figure 4.2). Two units of the oceanic plate stratigraphic succession are recognized in this district. The tuffaceous red shale and shale-dominant layers are generally significantly sheared and have the mélange facies. These shear structures are due to accretionary processes associated with the subduction of the oceanic plate [5].

The geological structure around the Tsubonouchi Area generally dips to the east or northeast. A northeast-southwest fold structure is recognized. The basalt

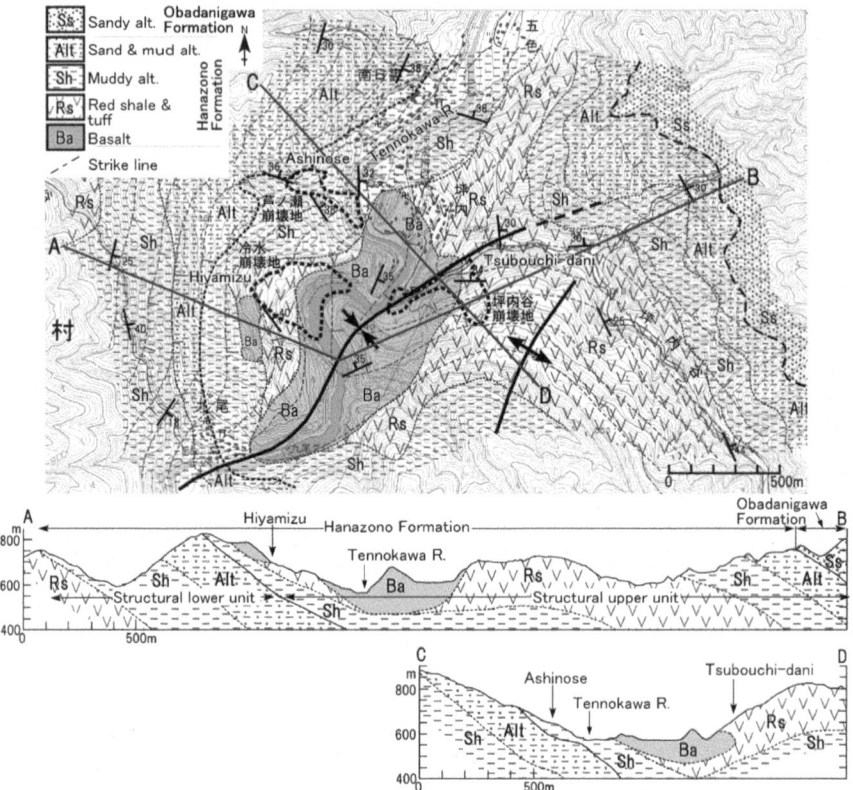

Figure 4.2 Geological map of Tsubonouchi Area, Tenkawa Village (using a topographic map "Minamihiura" at 1/25,000 scale by Geospatial Information Authority of Japan).

layers of the Hanazono Formation are distributed along the anticline axis, and many faults are recognized. The three large-scale slope failures are located in the basalt, basaltic tuff, and shale layers of the Hanazono Formation. Those locations are close to the boundary of the two sequence units. Although the slopes vary in direction, all slopes are generally close to the dip slope condition because of the fold structure. Because the former topography of these slope failures is characterized by slope variation landforms, slope destabilization may have occurred earlier due to the geological structure described above.

4.2.2 Nigoridani Area in Totsukawa Village

In the Nigoridani Area, the main area of slope failure was the valley slope from above sea level (a.s.l.) 900 m to a.s.l. 650 m, and the collapsed sediments flowed down as a debris flow (Figure 4.3). At a bend in the valley, part of the debris flowed over a small ridge. In this failure, a bore was generated by the

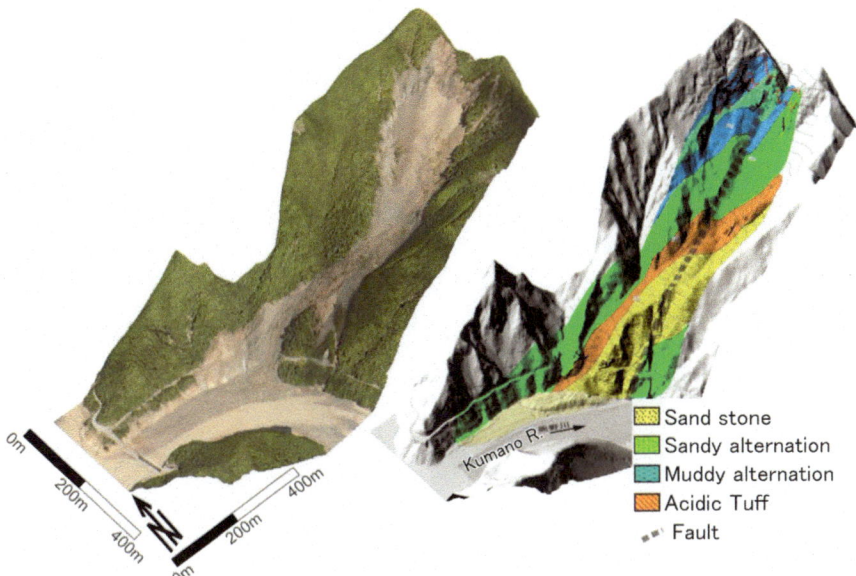

Figure 4.3 Overhead view and geologic projection of the Nigoridani site from the southwest. LP survey DEMs and photographs provided by Nara Prefecture were used.

debris flow that entered the main stream. The Nagatono Power Plant, located approximately 900 m upstream, was damaged by the bore propagation. This damage is described in a later section.

The Miyama accretionary complex of the Shimanto Terrane is distributed around the Nigoridani site [6]. Bedding, cleavage, and shear planes at this site generally dip northward at 60–70 degrees. The main part of the slope failure is mainly composed of alternating layers of sandstone and shale. The strata are generally severely fractured, and the sandstones are often blocky and lenticular (Photo 4.2). Shale also develops gently flexural cleavage and shear planes. Some east-west-oriented high-angle faults cutting these strata are recognized in the main scarp of this failure. Along these faults, there is significant deformation and rupture of the strata. Faults are also found in outcrops along the left bank of the main stream below the Nigoridani Valley, and the direction of the faults is consistent with the direction of the valley.

In the northern part of the main scarp, a smooth surface with fractures, which appeared to be a slip surface, was exposed. The fracture zone is about 2 m wide (Photo 4.3), and many gently bending fracture surfaces can be observed. The knickpoint of the shear face is browning, and a spring (dark area in the lower center of the photo) is observed.

Below a.s.l. (above sea level) 650 m of the Nigoridani, acidic tuff and sandstone beds are distributed, which are harder than the main part of the

Photo 4.2 Significantly fractured sandstone-shale alternation exposed on the main scarp in the Nigoridani slope failure site (Photo by Muneki Mitamura).

failure zone. Therefore, the sandstone-shale alternation, which was significantly deformed and disturbed, was destabilized and failed. The lower limit of collapse was regulated by the upper surface of hard bedrock consisting of acid tuff and sandstone. The collapsed sediment flowed as a debris flow along the lower part of the valley, which is composed of hard bedrock.

4.2.3 Nojiri Area in Totsukawa Village

The main failure area in the Nojiri site is from a.s.l. 780 m of the ridge to a.s.l. 550 m (Figure 4.4). In the southwestern part of the failure area, there is a slid block forming a small hill. Based on the topographic features before and after the failure, this block slid out from near the ridge crest. The failed sediment flowed down as debris flowed to the main stream of a.s.l. 210 m. The area around the Nojiri site is composed of sandstone-shale alternation and muddy mélange including basalt corresponding to the Ryujin accretionary complex [6] in the Shimanto Terrane. The main part of the failure area above a.s.l. 550 m consists mostly of muddy mélange layers. The main part of the failure area is recognized to have formed a dip slope on cleavage and shear planes. The sandstone-mudstone alternation is exposed near the crown of the failure area. This alternation is assumed to be in contact with the muddy mélange in the main part of the failure through a thrust.

Photo 4.3 Side scarp conditions and springs in the Nigoridani slope failure site (Photo by Muneki Mitamura).

The main failure area is an area subjected to crushing and shearing associated with fault movement and is fractured. This suggests a situation in which rainwater from upslope could easily infiltrate into the subsurface. On the other hand, a large block of basalt is located below the main part of the failure area at a.s.l. 450–550 m. Photo 4.4 shows a steep cliff formed by basalt blocks. The main part of the failure is located on the upper left side of this photo. The basalt block is a hard bedrock that forms a steep cliff within the failure site and appears to be a less permeable area than the surrounding rock body. This suggests that the basalt block at the lower end of the failure area intercepted the water, which caused the groundwater level to rise in the main part.

4.2.4 Factors of Large-Scale Slope Failures in Accretionary Complexes

Based on the research, the predisposing factors for large-scale failures can be summarized into three conditions: vulnerability, structural orientation, and hydrologic conditions (Figure 4.5). The occurrence of large-scale failures is

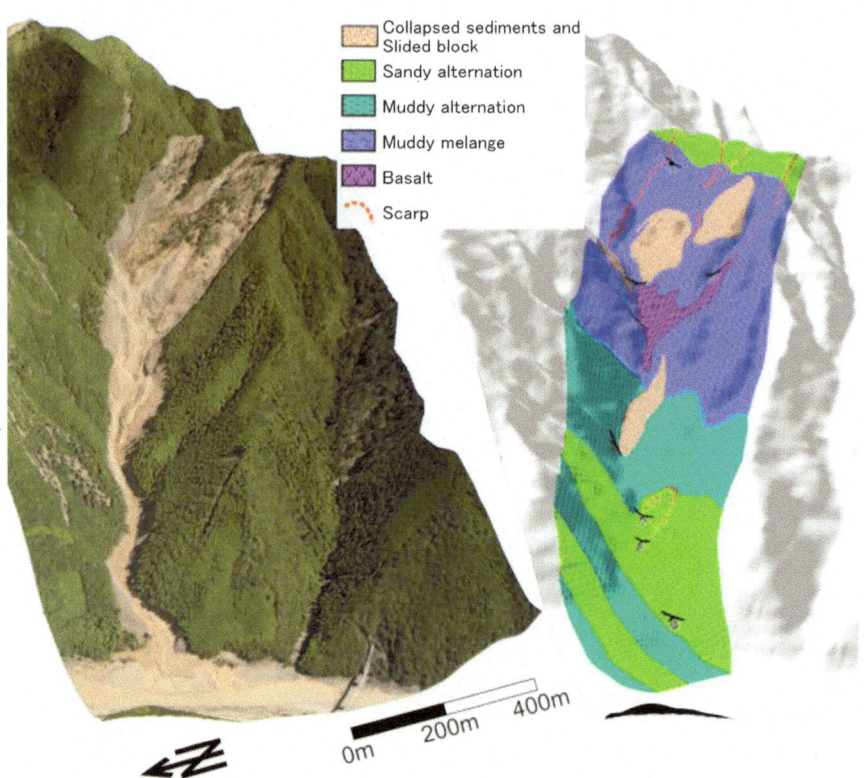

Figure 4.4 Overhead view and geologic projection of the Nojiri site from the south-west. LP survey DEMs and photographs provided by Nara Prefecture were used.

Photo 4.4 A large basalt block exposed just below the main part of the Nojiri slope failure site (Photo by Muneki Mitamura).

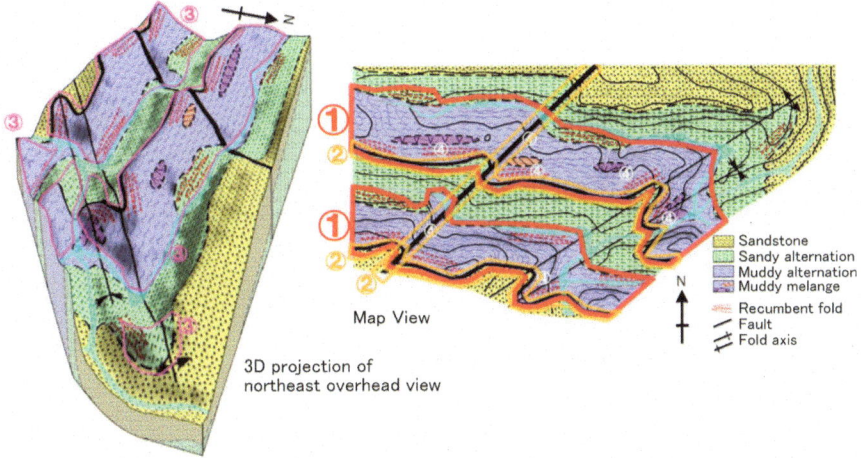

Map View

3D projection of
northeast overhead view

Sandstone
Sandy alternation
Muddy alternation
Muddy melange
Recumbent fold
Fault
Fold axis

Figure 4.5 Summary of the predisposition factors involved in large-scale slope failures
in accretionary complexes.
1. Conditions of Vulnerability: ①Structurally sheared stratification in
accretionary complexes ②Fracture zones of thrusts and transverse faults that
form accretionary complexes
2. Conditions for structural orientation to slip on a large scale: ③Bedding
planes, shear planes, and cleavage planes develop parallel to the slope (dip
slope condition)
3. Hydrological condition: Transverse faults and anticline structures that provide good permeability from the ground surface; Muddy rocks with many
fractures that act as groundwater stagnation zones; Chert, basalt, and thick
sandstone layers that induce groundwater damming up 4. Composite of factors.

not caused by these geological and hydrological conditions in isolation, but
by a combination of them.

4.2.4.1 Vulnerability

With the accretion process, the accretionary complex develops a pronounced
structural accumulation with thrust. The development of these thrusts has
created a sheared and fragile rock mass. In particular, the Hidakagawa
Terrane in the northern part of the Kii Peninsula has a history of reaching
deep into the crust and subsequently rising, and shear is prominent. The thick
brittle zone develops in the muddy areas located in the lower part of the
thrust sheet. In the terranes of accretionary complexes composing the
southern part of the Kii Peninsula, the degree of shear and deformation is
weaker to the south. However, since these terranes are also basically characterized by accretionary complex structures, the areas along thrusts are
significantly disturbed by geologic structures. In addition, there are areas of
weakness with significant shearing in the Neogene and later faults after the
formation of the accretionary complex and around the fold axes.

4.2.4.2 Structural Orientation

More than 60% of the large-scale slope failures were dip slope conditions in which foliations developed in the same direction as the slope inclination. Bedding, shear planes, and thrust fractures by sedimentary and accretion processes form the foliation in the rock mass. These dip slope conditions are a major factor in increasing the magnitude of failure. Not only this, but in many cases, faults since the Neogene period have high-angle fault plane form the lateral cliffs and mid-axis of the failure area. In the Otonashigawa and Muro Terranes, where shearing is less pronounced, it is regulated by the combination of bedding and their orthogonal joints. In addition, a major slope failure occurred at a location involving a fold structure with large deformation.

4.2.4.3 Hydrologic Conditions

It is clear that the major slope failures were primarily triggered by record-breaking heavy rains that allowed large amounts of water to infiltrate the slopes. Sheared shale rocks and thrusts developed in accretionary complexes, fracture zones high-angle faults, joint systems developed in sandstone beds, and open fissures formed by the dissolution of calcite veins, etc., facilitate the underground infiltration of rainwater. The large-scale slope failure area has undergone a precursory creep variation, the surface portion of the slope has loosened, and open cracks have developed significantly in the bedrock. Such conditions can cause rainwater infiltration to easily reach deeper into the bedrock (Photo 4.3, Figure 4.6). Significantly sheared and fractured shales are groundwater reservoirs. This progressive retention causes significant water pressure in the rock and slope instability. In cases where thick sandstone beds develop beneath the muddy rock, or where large chert/basalt hard bodies are contained within the muddy mélange, they act as impermeable zones. In situations where these rock bodies are located below the slip surface, at the lower end of the slope, or where an anticlinal structure develops, groundwater can easily be trapped in the slope composed of muddy mélange.

4.3 Surface Failure

In Nara Prefecture, more than 1,800 slope failures occurred, mainly in the southern area of the prefecture where heavy rainfall was concentrated. Large-scale slope failures accounted for only about 4% of slope failures in the prefecture, and most of failures were less than 10,000 m [2,8]. As shown in Photo 4.5, most of the slopes where surface failures occurred were 10 to 30 m wide and 50 to 100 m long. These slopes were relatively steep but not highly water-collecting slope geometries. The material making up the surface layer of the slope was relatively dominated by sandy gravel with low fine-grained matrixes.

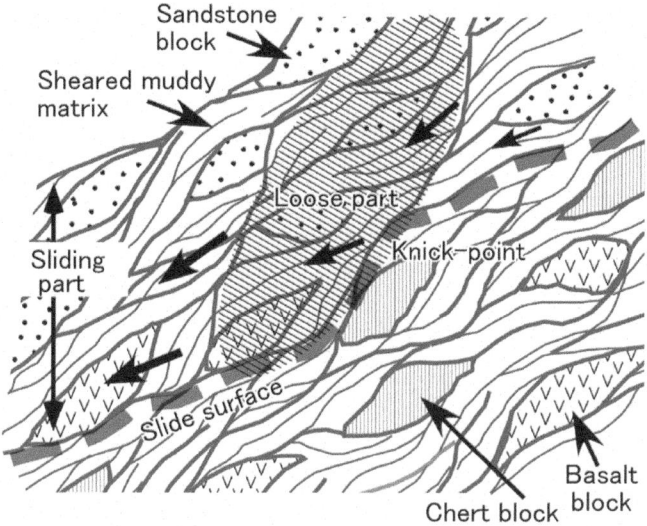

Figure 4.6 Slip surface and loosening area of bedrock in the mélange. The slip plane (a thick dashed line) is formed in the mélange consisting of lenticular broken blocks and sheared shale. On the knick-point where the slope of the slip surface increases (a shaded area), the loose zone in the bedrock allows groundwater from the surface to infiltrate more easily.

Photo 4.5 Situation of surface failure (Kotochi, Kamikitayama Village, Yoshino County) (Photo by Nobuyuki Torii).

4.3.1 Relationship with Rainfall

The distribution of failure sites and radar AMeDAS analyzed rainfall is shown in Figure 4.7. From this distribution, two characteristics of rainfall are found in areas where surface failures occur frequently. The first was characterized by slope saturation increased by preceding rainfall with a cumulative rainfall exceeding 450 mm, followed by areas with rainfall intensity exceeding 30 mm/h. Another feature is the location where the cumulative rainfall is more than 600 mm. Rainfall from Severe Tropical Storm Talas has a 30 years probability

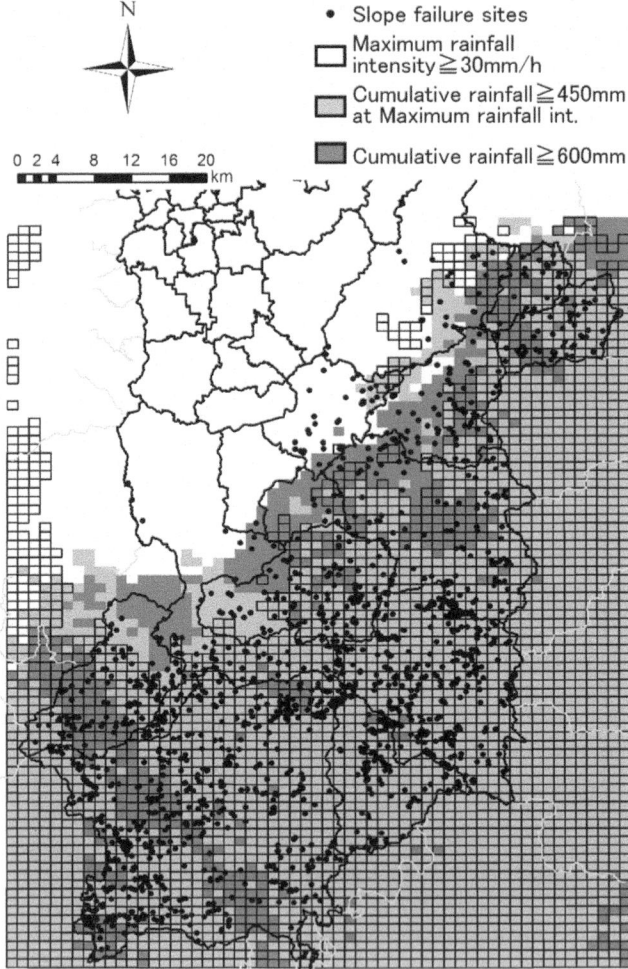

Figure 4.7 Distribution of cumulative rainfall exceeding 600 mm, maximum rainfall intensity exceeding 30 mm/h, cumulative rainfall exceeding 450 mm until the maximum observed rainfall is observed, and slope failure sites.

based on hourly and daily rainfall, but a 150–200 years probability based on 48–96 hours of rainfall. This suggests that the rainfall characteristics of longer duration and higher total rainfall greatly affected the distribution and morphology of surface failures. However, there was no tendency for surface failures to be concentrated in areas of particularly high rainfall intensity.

4.3.2 Topographic Features

Topographic features of the surface failure site were analyzed in GIS using a digital elevation model. The density of sites where surface failures occurred (failure density) was 0.62 sites/km^2, a relatively small value despite the high total rainfall. The slope angle of surface failures is predominantly 35 to 40 degrees (Figure 4.8). Slope orientation is mostly east (northeast to southwest, Figure 4.9). The classification of the slope shape into nine typologies showed that most of the failure slopes were of the convex/ridge type (① in Figure 4.10) and the straight/parallel type (⑤ in Figure 4.10) (Figure 4.11). The reasons for the large number of failures on non-catchment slopes are suggested as follows. Because southern Nara Prefecture is a high rainfall area, valley-shaped slopes had already failed due to past rainfall. In contrast, on slopes where the catchment area is small and rainwater is less concentrated, the soil making up the slope surface layer is not easily saturated by normal rainfall. Heavy rains that occur once every several decades with cumulative rainfall exceeding 600 mm are thought to have saturated the surface soil and caused surface failures.

4.3.3 Mechanism of Surface Failure

In order to investigate the mechanism of surface failure, field investigations, soil tests, and numerical analyses were conducted on a failure site in

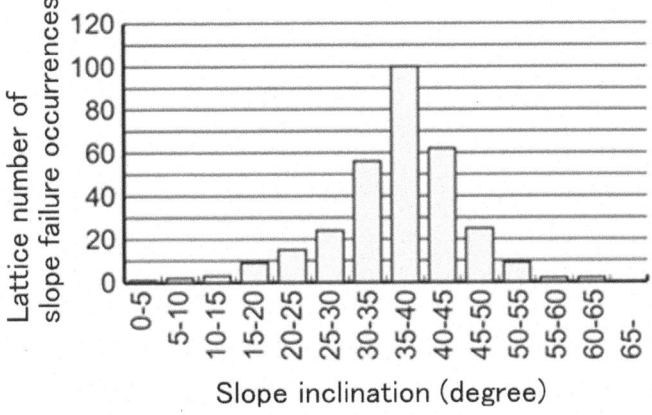

Figure 4.8 Distribution of slope angles at slope failure sites.

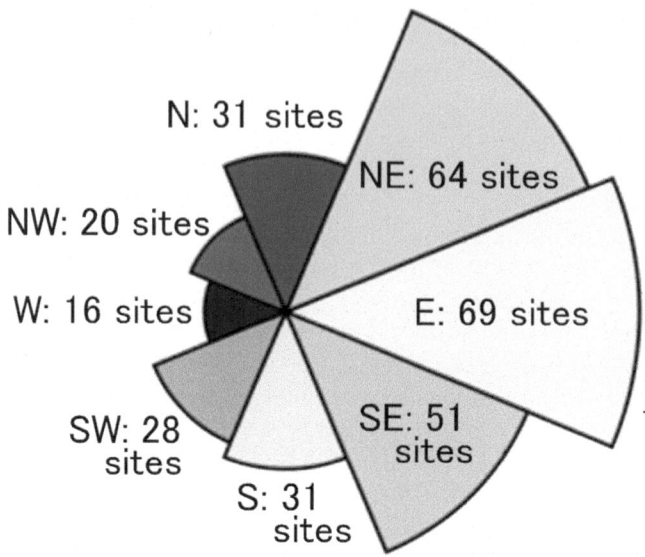

Figure 4.9 Distribution of slope azimuth at slope failure sites.

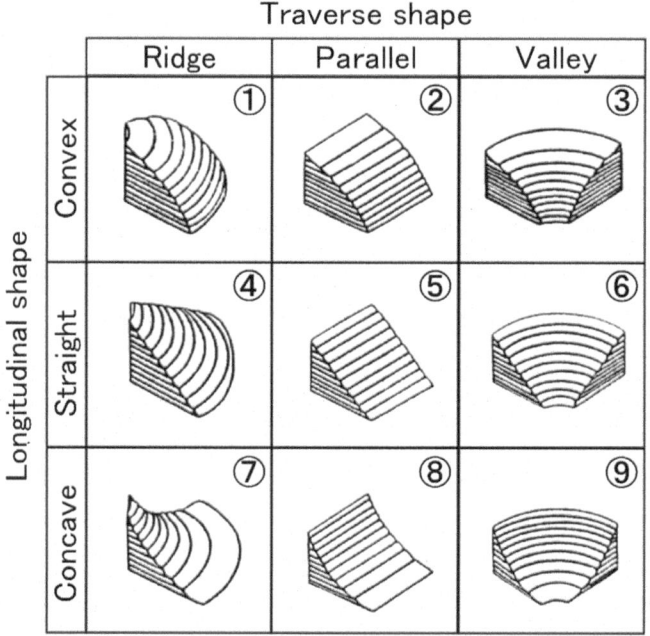

Figure 4.10 Typology of slope profile.

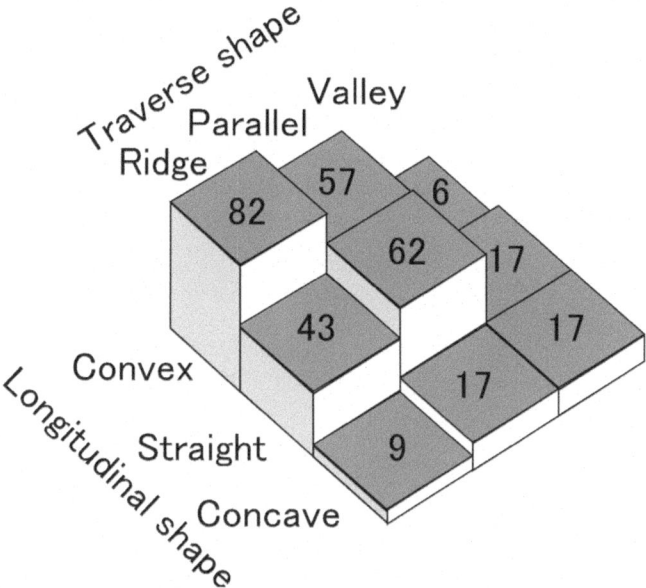

Figure 4.11 Distribution of slope profile at slope failure sites.

Nosegawa Village. This surface failure occurred on a slope with a 40-degree slope and a failure depth of about 1 meter. The results of the unsaturated seepage flow analysis are shown in Figure 4.12 and Figure 4.13. The saturation distribution (Figure 4.12) shows that a high saturation zone was formed on the entire slope at the time of the maximum rainfall intensity, and the saturation level was over 90%. The water table is located at the bottom of the slope as shown in Figure 4.13. The upper part of the slope where the failure was occurring (Figure 4.14) showed little water table formation. The results of stability analysis [9] using these results are shown in Figure 4.14 and Figure 4.15. The hazardous slip soil block, which represents the minimum factor of safety, coincides with the actual location of the surface failure (Figure 4.14). Rainfall and the change in the safety factor of this slope obtained from the analysis are shown in Figure 4.15. The safety factor of the slope drops sharply to 1.13 when the cumulative rainfall exceeds 136 mm. The analysis showed that the factor of safety was 1.09 when the cumulative rainfall exceeded 300 mm, and furthermore, it was 1.06 when the maximum rainfall intensity was recorded after the cumulative rainfall exceeded 600 mm. At this time, the entire slope has increased to about 90% saturation. The results of the analysis suggest that the surface failure was caused by the loss of stability due to the increase in self-weight and loss of apparent cohesion from high saturation.

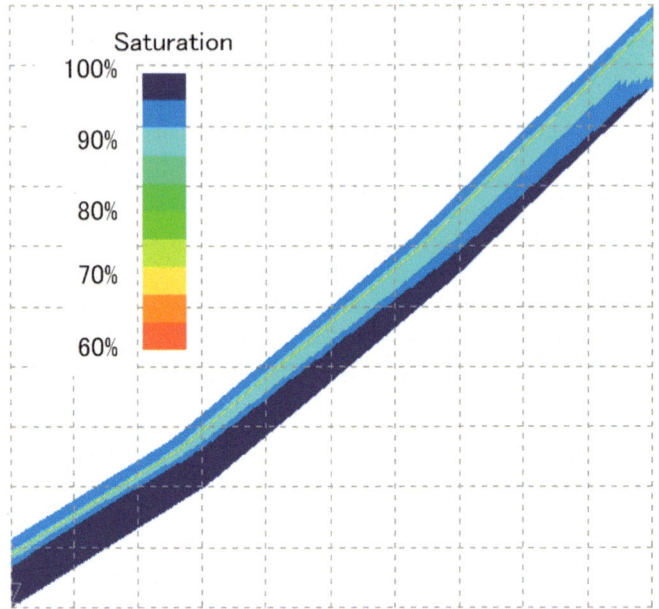

Figure 4.12 Saturation distribution map of seepage flow analysis results at the time of observation of maximum rainfall intensity.

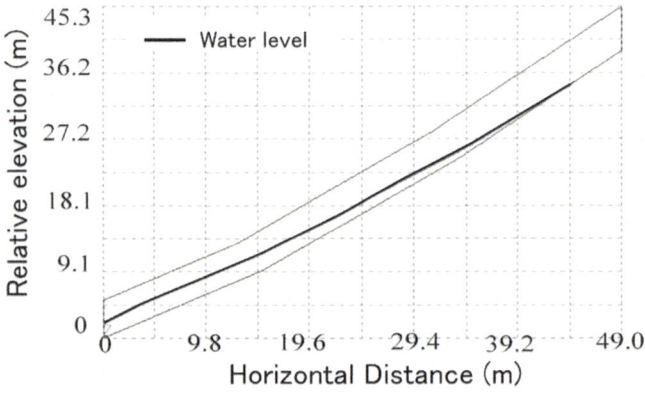

Figure 4.13 Groundwater level status at the time of observation of maximum rainfall intensity.

4.4 Damages along Rivers

The characteristics of riverine disasters in the Nara area are not those caused by floods due to heavy rainfall, but those caused by large-scale slope failures and the inflow of driftwood and sediment into the river channel as a result

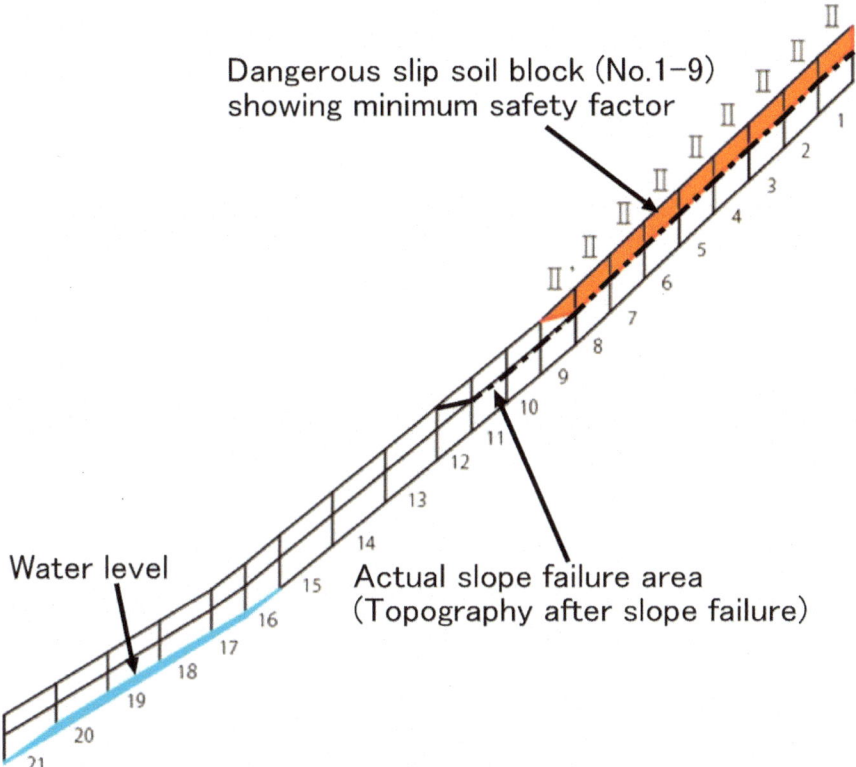

Figure 4.14 Comparison of hazardous slip soil block showing minimum safety factor, and actual failure location. The upper soil layer (2 m from the surface) is shown schematically.

of the failures. The Kumano River Basin was particularly heavily damaged. The Kumano River flows through the bottom of a mountain valley and has no embankments. There are revetments and bridges across the valley in places, and houses are located in the vicinity of the river. Damage includes scouring of revetment works and hinterland, falling bridges, burial due to sediment inflow, and flooding because river channels get blocked. Typical damage conditions are shown below.

4.4.1 Damage of Revetment Works and Hinterland

Photo 4.1 and Photo 4.6 show the damage in Tenkawa Village, the uppermost basin of the Kumano River. Three major slope failures occurred in the Tsubonouchi Area in this village. At Ashinose Valley site, the failed sediment flowed into the river channel as a debris flow. Due to the inflow of sediment, water overflowing from the river channel has scoured the opposite bank's

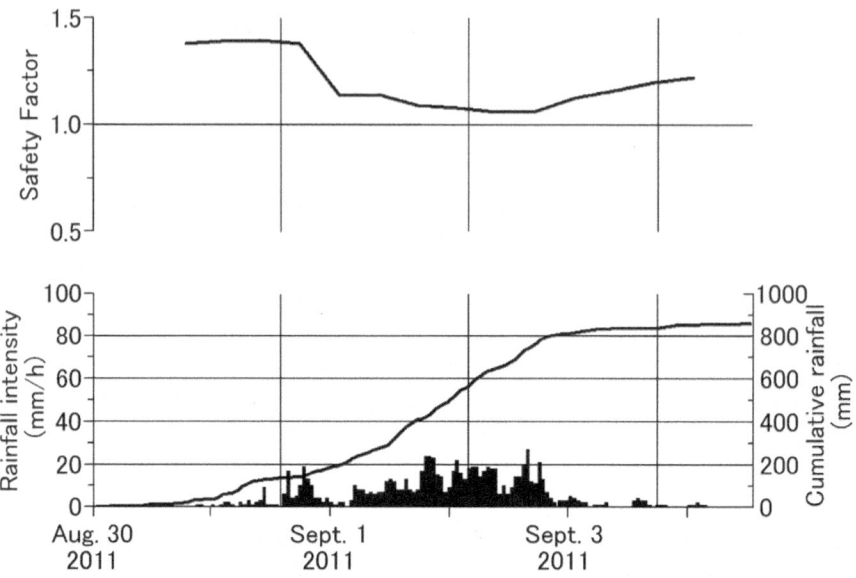

Figure 4.15 Relationship between transition in safety factor and cumulative rainfall.

Photo 4.6 Scouring revetment and playground of Tenkawa Junior High School
(Photo provided by Tenkawa Village)

revetment, school ground, and residential land. The lower part of Photo 4.6 shows the inflow area from Ashinose Valley, and the upper part shows the Tenkawa Junior High School ground and adjacent residential land, which were sharply cut away. A resident of this housing site died when his entire house was swept away. Without sediment inflow, the rainfall intensity was 20–30 mm/h, the same as the previous day, and the channel spacing capacity was maintained due to increased flow.

4.4.2 Damage of Bridges

The following is an example of damage to a bridge, the Oritate Bridge in Totsukawa Village. The Oritate Bridge is an upper truss bridge in the center and plate girder bridges on both banks. The damage seen from downstream is shown in Photo 4.7. The right-bank side plate girder bridge and the truss bridge in the center of the bridge fell down. Driftwood is hanging from the remaining lower chord of the truss bridge on the left bank, indicating that the river water has risen to this level. The cause of the bridge collapse is thought to have been the slope failure from the upper center of the photo, which caused a large amount of sediment and standing timber to flow into the river channel. The bridge-falling process is assumed to be as follows: The water level rose at the bend of the river channel, resulting in a small clearance with the water surface. Many pieces of driftwood got caught on the truss girders. This caused the truss girders to fall because of the increased flow of water pressure, which also caused the plate girder bridge to fall.

4.4.3 Sediment Inflow

The following is an example of the buried channel due to the sediment inflow in the downstream section of the Tubonouchi Valley slope failure site. The Tubonouchi Valley is the left branch of the Kumano River (Photo 4.1). This site is about 300 m from the confluence of the main river, and this section was

Photo 4.7 Collapsed Oritate Bridge in Totsukawa Village (photo taken from down-stream) (Photo by Koji Hirai).

Photo 4.8 Buried bridge at Tubonouchi-dani (taken from downstream) (Photo by Koji Hirai).

filled with debris flow. Photo 4.8 shows the buried bridge over the valley just before the confluence of the main river. What looks like a small waterfall in the center is the bridge railings. The debris flow was not limited to the river channel, and it buried a section of an adjacent shrine.

The most significant sediment inflow into the river channel was from the Ui (Shimizu) Area of Totsukawa Village to about 5 km downstream. In some areas, the riverbed elevation has risen more than 10 m, and sediment deposition is estimated at about 5 million m^3. The amount of sediment inflow is enormous and remains in the river channel because there is no place to dispose of the sediment after excavating the river channel. An example of partially implemented in-channel treatment of deposited sediments (shaping of embankment by pulling over one bank) is shown in Photo 4.9. However, since it is within the river channel, the upstream portion of the embankment on the left bank side is gradually being scoured, as shown in the photo.

4.4.4 Inundation due to River Channel Blockage

Photo 4.10 shows an example of flooding to the upstream residence area due to the blockage of the river channel caused by the inflow sediments from a large-scale slope failure. The photo shows that the floodwaters have

Photo 4.9 Sediment control measures in the river channel (Ohkuzure, Nagatono) (Photo by Koji Hirai).

Photo 4.10 Flooding in the Tsubonouchi Area (taken from upstream) (Photo provided by Tenkawa Village).

reached the roof level of houses. The location is the upstream area of the failure site of Hiyamizu, Tenkawa Village, shown in Photo 4.1. The failed sediment dammed the river channel for about four hours, causing the water level to rise to 20.7 m above the riverbed. During this period, the entire area at an elevation of 580 m, including Tenkawa Junior High School, was submerged. The backwater impact of the damming extended approximately 2.2 km upstream. Many of the submerged wooden houses were completely destroyed once they floated up. Characteristically, there is no sign of rainfall in the photographs. This is because deep failures occurred after the typhoon rains ceased. Even though the peak of the flood had passed and runoff was decreasing, the damming caused more damage.

4.4.5 Peculiar Example

Finally, the most unusual disaster cases are shown below. This is an example of damage to the Nagatono Power Plant in Totsukawa Village. The Nagatono Power Plant is a hydroelectric power plant consisting of a water conduction tunnel and power-generation facilities, as shown in Photo 4.11. National Road 168 runs behind the power station, with a steel tower on the left and houses on the upper part of the road on the right. After the disaster, as shown in Photo 4.12, the power station and houses have disappeared, and

Photo 4.11 Nagatono Power Plant before the disaster (Photo by Koji Hirai).

Photo 4.12 Nagatono Power Plant after the disaster (Photo by Koji Hirai).

the tower is bent. The cause of the damage is interpreted as follows, based on the fact that the direction of the tower bend is upstream and the condition of scouring on the slope. First, a slope at 1.5 km downstream of the power plant failed, damming the Kumano River. After the water level rose due to this damming, a failure in the Nigoridani Area about 1 km downstream caused a large amount of sediment to flow into the river channel. As a result, the water accumulated in the river channel and flowed back upstream as a bore, hitting the power plant. This generated a bore wave, which flowed back upstream and hit the power plant.

4.5 Summary

Field investigations of several large-scale slope failures in the Nara Prefecture area have revealed that the geological characteristics of accretionary complexes play a significant role in predisposing them to failure. Many large failures have occurred with the formation of a slip surface in the significantly fractured area that constitutes the lower part of the thrust sheet. Large slips occur on slopes with dip slope conditions where the separation plane— consisting of bedding planes, shear planes, and cleavage planes within the rock—is generally coincident with the direction of the slope. In addition, new faults and near fold axes that arose after accretionary complex formation also weaken the rock and contribute to slippage. Within these fractured rock bodies, rainwater can easily infiltrate deeper into the slope and groundwater can easily stagnate, leading to slope instability from deeper areas. Large blocks of chert, basalt, and other rocks in the accretionary complex and thick sandstone layers act as impermeable zones and tend to dam up groundwater. The above geological and hydrological conditions did not occur in isolation, but in combination.

The characteristics of the surface failures that occurred in the accretionary complex area of the Nara Region can be summarized as follows. In the accretionary complex distribution area in the southern part of Nara Prefecture, continuous rainfall from Severe Tropical Storm Talas reduced the safety factor by causing high saturation of the slope surface layers with slope geometries (ridge or parallel type) that are not prone to failure during rainfall. Under these conditions, further continuous rainfall brought cumulative rainfall in excess of 450–600 mm. Furthermore, on slopes subjected to strong rainfall with temporal intensities exceeding 30 mm/h, the increase in self-weight and the loss of apparent adhesion led to the occurrence of failures.

The remainder of this chapter presents typical forms of river disaster damage. These disasters occurred between September 3 and 4, when a major slope failure occurred. Considering that the same intensity of rainfall had occurred on September 2, the previous day, it is likely that the cause of the river-related disaster was the inflow of huge amounts of sediment and timber into the river channel due to slope failures.

References

[1] Nara Prefecture, 'Records of the 2011 Great Floods in the Kii Peninsula', p.14, 2013. (in Japanese)
[2] Deep-seated Slope Failure Research Group, 'Present status report on clarification of the mechanism of deep-seated Slope Failure at the 2011 Great Floods in the Kii Peninsula', *Nara Prefecture*, p.40, 2013. (in Japanese)
[3] Geotechnical Engineering Society of Japan Kansai Branch, Kansai Branch of the Japan Society of Applied Geology, Kansai Geological Surveyors Association, and Chubu Geological Surveyors Association: Report of the Research Committee on Response to Ground Disasters Caused by "Unexpected" *Heavy Rainfall*, p.404, 2015. (in Japanese)
[4] Based on information provided by Nara Prefecture (in Japanese)
[5] The Geological Society of Japan: *Regional Geology in Japan – Kinki District-*, Asakura Publishing Co., Ltd., p.448, 1999. (in Japanese)
[6] Kishu Shimanto Research Group: New Developments in the Study of the Shimanto Accretionary Complex in the Kii Peninsula, *Monograph of the Association for the Geological Collaboration in Japan*, No. 59, p.295, 2012. (in Japanese)
[7] National Research Institute for Earth Science and Disaster Prevention, 1/50,000 Landslide Topographic Distribution Map 'Sanjogatake', Vol. 23, 'Wakayama/Tanabe', 2005. (in Japanese)

[8] Ueda, Y., Irikuci, K., Yamagami, S., Matsuyama, S., Takahashi, T., Hirano, M., Simada, T., Hori, T., Ekawa, M., Takeshima, A., The Meiji Totsukawa Disaster and the Sediment Disaster Caused by Severe Tropical Storm Talas in 2011, Annual Conference 2012 Summary Collection of Japan Society of Erosion Control Engineering, pp.24–25, 2012. (in Japanese)

[9] Fumiiwa, H., Torii, N., Kato, M., Koizumi, K., Kagamihara, S., Matsumoto, S., Mitamura, M., Shibuya, S., Kawabata, M., A Mechanism of Slope Failure Due to Heavy Rainfall of the Typhoon No.12 in Nosegawa-Mura, Nara in 2011 Based on Saturated-Unsaturated Seepage Analysis and Slope Stability Analysis, Proceedings of the 50th Geotechnical Engineering Conference, pp.2279–2280, 2015. (in Japanese)

5 Disasters in Wakayama Prefecture

Yasuyuki Nabeshima

5.1 Introduction

Most of the geotechnical disasters caused by Severe Tropical Storm Talas (1112) in 2011 occurred in Kinan Region, which is located in the southern part of Wakayama Prefecture. They were characterized by large-scale slope failures, including deep slope failure in the Hidaka and Nishimuro Areas where accretionary prisms were distributed, and surface failures and debris flows in the Higashimuro Area, where both igneous and sedimentary rocks were distributed. Failures of river dikes have been confirmed at 11 sites in six rivers, and more than 1,000 flood disasters including damage to revetments have occurred. In addition, precious local resources, such as the Kumano Nachi Taisha Shrine and the Kumano Kodo ancient road, were severely damaged. Therefore, the characteristics of five types of geotechnical disasters in Wakayama Prefecture are described in this chapter: Large-scale slope failures, surface failures, debris flows, river disasters, and cultural heritage damage.

5.2 Geological Characteristics in the Southern Part of Wakayama Prefecture [1,2]

Figure 5.1 shows the geology and geotechnical disaster sites in the southern part of Wakayama Prefecture (Kinan Region); this figure only shows the disaster sites that were investigated by our research group. The geology in Kinan Region consists of mainly three types, such as the Shimanto Terranes that include Hidakagawa, Otonashi, and Muro Groups, the Kumano acid igneous rocks, and the Kumano-Tanabe Groups.

The Shimanto Terranes make up a major part of the Kii Peninsula, which was formed by the addition of sediments on the oceanic plate to the continental plate when the oceanic plate sank below the continental plate. As the oceanic plate sank below the continental plate, the sediments on the oceanic plate were severely deformed and complex geological structures were made, such as the formation of mixed rocks with various kinds of rocks and many large-scale reverse faults in which the upper plate was lifted. The Shimanto Terranes are classified into the Hidakagawa Group (70–60 million years ago), the Otonashi Group (60–50 million years ago), and the Muro Group

DOI: 10.1201/9781003375210-5

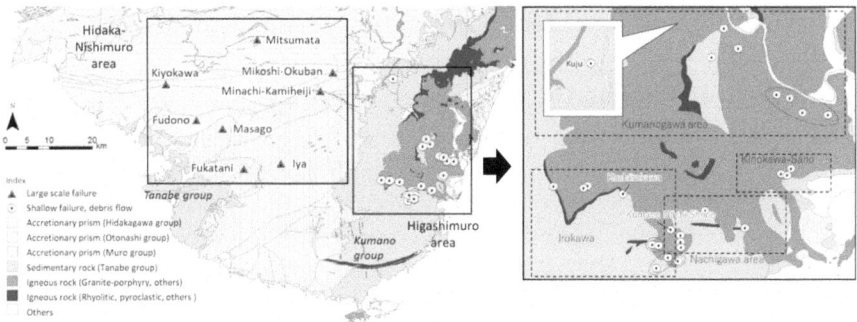

Figure 5.1 Geology and geotechnical disaster sites in the southern part of Wakayama
Prefecture.

(50–20 million years ago) from oldest to newest. These were significantly deformed during their formation, and thrusts were formed between geological
boundaries. Large-scale slope failures in the Hidaka-Nishimuro Area occurred
at joint planes, geologically disturbed structures, faults, and folds that occurred
in the Miocene.

The Kumano acid igneous rocks are magmatic igneous rocks, which are
mainly granite porphyry rocks and distributed in the southeastern part of the
Kii Peninsula. The columnar joints were formed in the granite porphyry rock,
which were formed by volumetric shrinkage during cooling time of the
magma. The weathering of rocks developed along the cracks between these
joints and proceeded from outside to center like an onion shape; finally,
weathered core stones with round shape were formed. The eruptions of
rhyolitic pyroclastic rock and dikes of rhyolitic and rhyolitic pyroclastic rocks
are distributed in this area. The rhyolitic pyroclastic rocks are mainly massive
forms, but in some places they contained the Kumano Group. They sometimes form smooth and flat surfaces. The granite porphyry and quartz porphyry rocks are distributed near Kuju Area (Kumanogawa, Shingu City)
along the Kitayama River, which are part of the Omine granitoids, and the
Shionomisaki igneous complex rocks (Basaltic, rhyolitic, and canalitic rocks'
to basaltic, rhyolitic rocks and gabbro) are distributed in Shionomisaki and
Oshima (Kushimoto town) at the southernmost end of the Kii Peninsula.

The Kumano and Tanabe Group consist of mixture such as gravel,
sandstone, and mudstone that were deposited on the continental shelf and the
continental shelf slope (i.e., forearc basin deposits). The primary stratigraphy
of the strata of these sedimentary rocks was preserved in good formation, and
some muddy breccia formed by mud diapir can be observed, but there was
little disturbance in the rock body such as accretionary prism. The incline of
the slopes in these areas is not steeper than that of the Kumano acid igneous
rocks. The Kumano Group is the name of a group of sedimentary rocks

distributed in the Higashimuro Area, and the Tanabe Group is the name of a group of sedimentary rocks distributed in the Nishimuro Area.

5.3 Characteristics of Rainfall That Caused Geotechnical Disasters

Figure 5.2 shows the relationship between accumulated rainfall and hourly rainfall in the event of a geotechnical disaster [3]. Radar AMeDAS rainfall is analyzed in this figure. Accumulated rainfall in this chapter is the total amount of rainfall from August 30 to 31, which was at the beginning of rain due to Severe Tropical Storm Talas and the occurrence period of slope failures. Hourly rainfall refers to the hourly amount of rainfall until the slope failures occurred. The figure shows that the hourly rainfall was 2 to 40 mm/h and the accumulated rainfall was 760 to 1,450 mm when the slope failure occurred in the Hidaka-Nishimuro Area, and the hourly rainfall was 40 to 120 mm/h and the accumulated rainfall was 720 to 1,330 mm when the slope failure occurred in the Higashimuro Area. Although the hourly rainfall in the Higashimuro Area was higher than that in the Hidaka-Nishimuro Area, there is no significant difference in the accumulated rainfall between the two areas. However, the rainfall characteristics of four sites shown as Kuju, Kashihara, Kuchiirokawa-Fukuchidani, and Onokawa in the Higashimuro Area were clearly different from other sites. Therefore, excepting these four sites, the hourly rainfall in the Higashimuro Area was 70 to 120 mm/h, and the

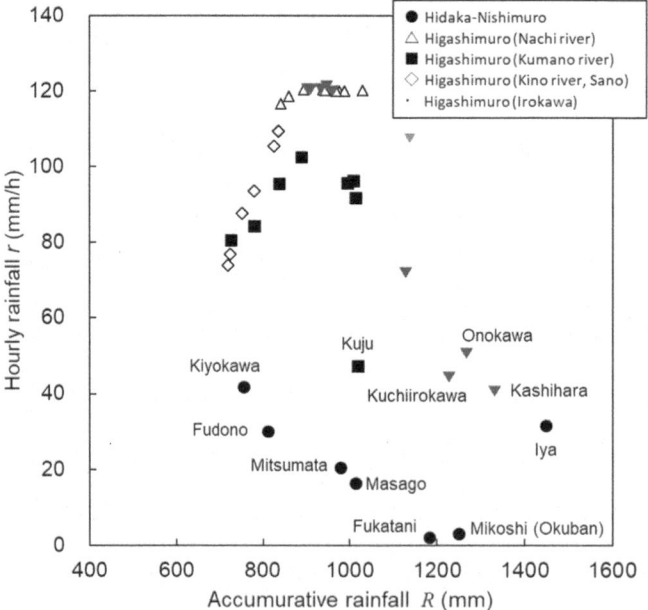

Figure 5.2 Relationship between accumulated rainfall and hourly rainfall in the geotechnical disasters.

accumulated rainfall was 720 to 1,130 mm. Although the total amount of rainfall in many sites in the Higashimuro Area until the sediment disasters occurred was less than that in the Hidaka and Nishimuro Areas, it can be seen that very heavy rain fell when the sediment disasters occurred.

In general, the relationship between sediment disaster types and rainfall amounts can be considered as follows: Large-scale slope failures including deep-seated landslide occur when rainfall duration is long and total rainfall is large; surface failures occur when rainfall amount in a short time is large, and debris flows occur when rainfall amount in a short time is large after total rainfall becomes large. Except for the four sites of Kuju, Kashihara, Kuchiirokawa-Fukuchidani, and Ohnokawa in the Higashimuro Area, the relationship between sediment disaster pattern and rainfall amount agreed well with the general trend in both the Hidaka and Nishimuro Areas and the Higashimuro Area.

Although the hourly rainfalls in Kuju, Kashihara, Kuchiirokawa-Fukuchidani, and Ohnokawa did not exceed 60 mm/h, accumulated rainfalls exceeded 1,000 mm when the collapse occurred. In other words, the surface failures and debris flows occurred at these sites because the accumulative rainfall became large; they were not caused by the short-duration strong rainfall after the total rainfall became large.

In conclusion, it is important when considering possible geotechnical disasters to take into account both accumulated rainfall and hourly rainfalls, such as large-scale slope failures in the accretionary prism and surface failures and debris flows in the igneous and sedimentary rocks. They are typical geological types in the Kinan Region in the southern part of Wakayama Prefecture. However, it is necessary to pay attention for irregular cases such as four sites in the Higashimuro Area. The surface failures and debris flows occurred in cases where the accumulated rainfall increased even if the hourly rainfall was not large.

5.4 Large-Scale Slope Failures Occurred in Accretionary Prism

5.4.1 Topography and Geological Characteristics of the Collapsed Area

Table 5.1 shows the characteristics of the large-scale slope failure sites investigated in the field study [4]. Six among the eight sites such as Mitsumata, Mikoshi, Okuban, Minachi, Kamiheijigawa, Manago, Fukatani, and Iya showed dip slopes in the main geological structures of stratigraphic and shear planes. These sites have common features such as the geological aspect (mainly composed of mudstone), collapsed slopes in the past, and the topographical aspect of convexity slopes, and also most of them had some faults across the collapsed areas. The Minachi has the peculiarity that dikes and hydrothermal alteration zones of quartz porphyry were distributed like dip slope, but others showed the above aspects. Two sites of Kiyokawa and Fudono had different topographical and geological aspects from other districts, characterized by high-angle stratigraphic surface with orthogonal direction with the failure surface, small water catchment areas, no collapse records, linear valleys, and/or small concave topography. These two failure

Table 5.1 Characteristics of large-scale slope failures in the Hidaka-Nishimuro Area

Failure Site	Geology	Rock Type	Geological Structure	Rainfall	Topography	Water Catchment Boundary	History of Failure	Failure Factors and Causes	Effect of Fault
Mitsmata	Hidakagawa group-Rujin	Shale >> Green rocks	I: 35–54°	B	Convex ridge, Clear valley slope	150 m	Old-colluvial soil	Erosion of valley slope and shallow area, Shallow failure of weathered rock and expand of failure area, Concentration of surface water	Fault crushing belt in about 2 m width under main scrap
Kiyokawa west	Otonashi group-Lower Haroku	Lutaceous alternation	III: 70–90° Vertical to failure surface	A	Uniform slope, Valley slope	10–50 m	No-failure history	Toppling failure, Rainfall water supply due to high angle structure	Fault across end of failure area and spring water
Kiyokawa east		Lutaceous alternation	IV: 30° Opposite slope	A	Convex, Ridge	100 m	Old-failure site	River channel erosion of lower end, Secondary failure of old-colluvial soil	
Fudono	Otonashi group-Haroku/ Uridani	Lutaceous alternation	III: High angle Vertical to failure surface	A	Ridge, Uniform depression (lower end)	10 m	No-failure history	Fault and high angle failure, Rainfall water supply,	Failure of Furuyadani thrust
Mikoshi-Okuban	Otonashi group-Lower Haroku	Lutaceous alternation	I: Bedding 60° I: Fault 60°	C	Convex ridge (slight depression at lower area)	0–20 m	Failurein 1989 & 1953	Dip slope failure along thrust, Groundwater supply, River channel erosion of lower end	Failure of Hariyasu thrust in structural unit

Site	Group	Rock type	Type / Angle	Class	Topography	Elevation	Soil	Failure description	Fault
Minachi-Kamih-eijigawa	Otonashi group-Lower Haroku	Lutaceous alternation, Hydrothermal alteration, Quartz porphyry	I: 35–44° Dikealteration, Lutaceous alternation	unknown	Convex ridge (depression)	0 m	Thick old-colluvial soil	Collapse of dikes, alteration dikes, and mudstone. High-angle faults at both north and south ends of the collapse site. Water barrier by fault	Neighborhood thrust. Fault across failure area (88°)
Masago	Muro group-Upper Nishitani	Sandstone, Conglomerate, Thin-mudstone	I: 35–40°	B	Lower slope of convex ridge	150–200 m	Thick old-colluvial soil	Dip slope failure and secondary failure of old-colluvial soil, Water barrier by fault	Fault across failure area
Fukatani	Muro group-Uchikoshi	Lutaceous alternation	II Bedding 10° I Fault 45–60°	C	Uniform depression (about 100 m width)	700 m	Old-debris flow or old-colluvial soil	Dip slope failure and secondary failure of old-colluvial soil, Water barrier by fault	Fault across failure area

Type I: Dip slope A: Small accumulative rainfall and large hourly rainfall.
Type II: Moderate dip slope B: Medium accumulative rainfall and medium hourly rainfall.
Type III: Moderate opposite slope C: Large accumulative rainfall and small hourly rainfall.
Type IV: Opposite slope.

sites had similar features in that they were located on thrusts or faults described in historical geological maps.

In the current field study, the main topographic and geological aspects found in the large-scale slope failure sites in the area are: (1) the geological structures were dip slopes, (2) the geology is mainly composed of mudstone, (3) there were collapsed slopes in the past, (4) their topographical aspect was of convexity slopes, and (5) the faults crossed the collapse sites. These are corresponding factors for large-scale slope failures regardless of the geology. In addition, the topographical and geological aspects, such as the vicinity of the thrust or the attack slope of the river, observed at the slopes, make the risk of large-scale slope failure potentially high. For predicting large-scale slope failure in the future, the features (1) to (4) can be narrowed down to some extent by historical literature and 1:25,000-level topographical maps and aerial photographs. However, the features in (5) may have significantly affected the groundwater level, which is the main trigger of the collapse, but most of these are small-scale features that are not described in the existing geological maps. In order to identify a small fault, steady investigation such as site surveying is necessary, and it is a difficult item to take up in a collapse prediction investigation.

5.4.2 Large-Scale Slope Failures with Geological Structure Like Dip Slope

The Kumano District is a typical example of a large-scale slope failure with geological structure like dip slope. A large-scale slope failure of about 400 m in width and 280 m in height occurred in the Kumano Area of the Hiki River Basin. The geology of the collapsed area was composed of thick sandstone and mudstone alternate layers of the Muro Group in Uchigoshi Formation and the mudstone of the Kogawa Formation which overlapped with it, and this caused the dip slope failure. In the geological map of Suzuki et al. [5], a fault is described near the boundary between the Uchikoshi Formation and the Kogawa Formation. Since this fault is a dip slope and an important predisposing factor for collapse, detailed field surveys were carried out to understand the geological conditions and properties of the fault. As a result of the investigation, three factors were considered to cause the collapse: (1) the distribution of sedimentary rocks in the form of dip slope, (2) the distribution of old collapsed areas, and (3) the distribution of dip slope in the fault fracture zone with low shear strength, and the triggers of the collapse were (1) the infiltration of rainwater into the ground by Severe Tropical Storm Talas and (2) the rise of pore water pressure in the ground. Water supply from Hyakkendani (Hyakkendani) on the back side of the double ridges is the most likely route for rainwater infiltration and groundwater supply [6,7]. Although continuous outcrops without cracks are distributed in the riverbed of Hyakkendani, the supply of groundwater along the fault fracture zone is considered because it is relatively permeable. The presence of unstable topography on the eastern slope of the collapse site, which may have been a former collapse site, suggests that groundwater supply along the fault and sliding of the ground occurred over a wide area including the collapse site even before the collapse.

5.4.3 *Large-Scale Slope Failure without Geological Structure Like Dip Slope*

As described in Table 5.1, in the Fudono Area of the Otonashigawa Group, there was no collapse history and landslide terrain, and the collapse occurred without the dip slope. The collapse was about 200 m of east and west, and about 100 m of north and south, reaching a specific height of about 120 m, and the end of the collapse extended to a distance of about 420 m from the top. This site has not been reported as having a landslide that occurred in the past, and typical landslide topography could not be found from topographical maps and aerial photographs before the collapse, and the top of the collapsed site showed ridge topography. The geology of the collapse is the Uritani Formation and the Hamoku Lower Sandstone Formation of the Otonashigawa Group, and the Furuyadani Thrust passes along the collapse as shown in Figure 5.3. The geological structure near the collapse site is very complex because of bending deformation. The geological structure in the collapsed area was steeply inclined and across nearly perpendicular to the collapsed area, and piping holes were confirmed in the collapsed area, suggesting that highly permeable areas were locally formed with the development of small shear bands and joints. On the other hand, focusing on the surface layer on the upper part of the foundation, the water retention capacity was probably small because the layer thickness was less than 1 m. Based on these findings, it can be pointed out that rainwater penetrated the surface layer and infiltrated along these structures when a large amount of rain fell, and when the infiltrated rainwater became a continuous water vein from the surface, the high water pressure pushed the soil mass to the front and caused the large collapse [8].

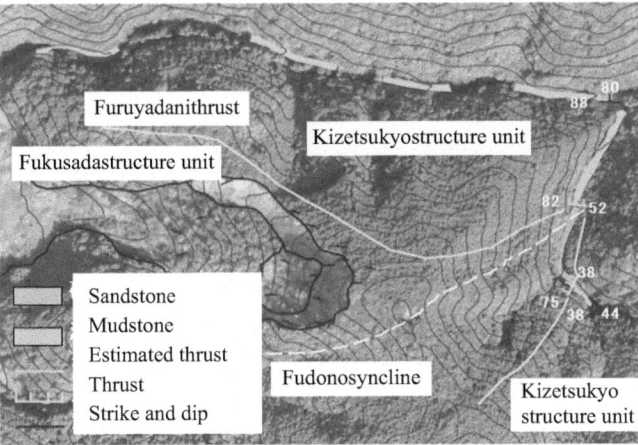

Figure 5.3 Geology in the Fudono Area and location of Furuyadani Thrust.

5.5 Surface Failures and Debris Flow in Igneous and Sedimentary Rocks

In the Higashimuro Area, surface failures and debris flows occurred frequently in the Kumano River Basin (Kumanogawa Town, Shingu City), Kinokawa and Sano (Shingu City), Nachi River Basin (Nachikatsuura Town) and Irokawa River (Nachikatsuura Town). The geology of these areas was composed of the Kumano acid igneous rocks (granite porphyry, rhyolite, rhyolitic pyroclastic rock) of igneous rocks and the Kumano group of sedimentary rocks, and the collapses were characterized by their geologic groups. Here, the characteristics of sediment disasters are described by dividing them into Kumano acid igneous rocks and the Kumano formation group.

5.5.1 *Kumano Acid Igneous Rocks [9–11]*

Granite porphyry, which is a main rock of the Kumano acid igneous rocks, develops columnar joints and is easily weathered along the cracks of these joints (Photo 5.1). The distributed area of granite porphyry is steeper than that of the Kumano Group, while the Kumano Group forms a gentle incline slope. Reflecting such topographical and geological conditions, most of surface failures in this area occurred in the distribution area of granite porphyry. In these collapsed areas, collapsed soil (which was the old debris flow

Photo 5.1 Expose of granite porphyry along Kumano River in Mie Prefecture (Photo by Haruhiko Yano).

deposit) was deposited on the gentle incline slopes of the river, which caused debris flow with the sediment due to the surface failures. Also, the collapsed soil remaining in the rivers collapsed again and caused the debris flow.

Figure 5.1 shows the geological condition of the Nachi River Basin, which illustrates that many debris flows occurred at the mountainous sedimentary areas in the mountain torrents, which are assumed to be the old debris flow deposits. In other words, in the mountain torrents where debris flows occurred in S.T.S. Talas (1112), it was supposed that debris flows occurred in the past and the debris remained as avalanche soils in the gentle incline slopes of the torrents.

The main rock in the Kumano acid igneous rocks was a granite porphyry, but there were some rhyolitic pyroclastic ejecta and dikes of rhyolitic and rhyolitic pyroclastic rocks. Surface failures and debris flows that were origi-nated from rhyolitic pyroclastic rock occurred in Hasegawa in the Nachi River Basin and Kuchiirokawa, Akahatadani, and Kashihara in the Irokawa Area.

Around middle Hase River, a relatively large-scale collapse occurred in a tributary on the left bank, where the rhyolitic pyroclastic rock area was. Avalanche soils on the hillside slope collapsed; the geology of the collapse origin was not rhyolitic pyroclastic rock but granite porphyry in a seamless geological map. Although some avalanche soils partially remained on the slope, flat and hard bedrocks were exposed under the slope (Photo 5.2). Rhyolitic pyroclastic rocks were mainly massive rocks and sometimes formed

Photo 5.2 Failure site of the left bank around middle Hasegawa River taken in 2012 (Photo by Nobuyuki Egusa).

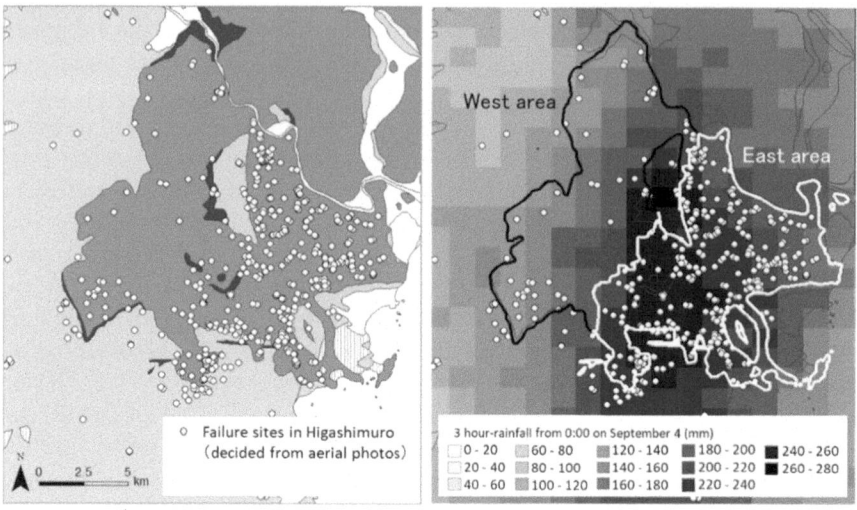

Figure 5.4 Collapsed sites in the Higashimuro Area (Right: Geological condition; Left: three-hour rainfall).

flat bedrock. That is, the collapse site in the Hase River Basin showed a typical geology of rhyolitic pyroclastic rocks, where the avalanche soil flowed out and exposed the bedrock.

Figure 5.4 shows the collapsed sites (434 sites in total) in the Higashimuro Area plotted on a seamless geological map, which were extracted from an orthologized aerial photograph provided by Wakayama Prefecture. Judging from this figure, many collapses were concentrated in the eastern area of the Kumano acid igneous rocks. The density of collapse in the eastern area at 5.4 sites/km^2 was about seven times wider than that in the western area at 0.8 sites/km^2. Through a comparison of predisposing factors, such as slope angle, slope orientation, slope shape, and vegetation, there was no significant difference between both areas. Instead, the steeper slope angles of more than 35 degrees were more often in the west area, but there were fewer collapse sites. Photo 5.3 shows the three-hour rainfall distribution (using Radar-AMeDAS analysis) from midnight to 3:00 a.m. on September 4. It clearly shows that the densities of the collapses in both areas resulted from the amount of precipitation just before the collapse. From these results and Figure 5.2, although the Higashimuro Area suffered severe damage due to frequent large-scale debris flows, a major factor was the strong rainfall of more than 70 to 80 mm (220 mm in three hours of rainfall), when the deposit soils loosened by accumulated rainfall of nearly 700 mm.

5.5.2 *Kumano Group*

The number of sediment disasters that occurred in the Kumano Group were fewer than the Kumano acid igneous rocks, and the scale of collapses was

Photo 5.3 Mud rock in Kumano Group with many joints and dip slope structure (Photo by Nobuyuki Egusa).

also smaller. Although the number was a few, the igneous rock bodies and igneous dikes were commonly distributed behind the collapse slopes at Kuju and Kuchitakata in the Kumano River Basin (Figure 5.1), and the collapsed slopes showed a dip slope with stratigraphic streambeds (Photo 5.3). The igneous rocks in this area were formed by magma that penetrated accretions and sedimentary rocks due to magmatic activity 15 million to 14 million years ago. These geological processes usually affected the development of joints in the mudstone dominant layers of the Kumano Group, which are located close to the Kumano acid igneous rocks and the Omine granitic rocks. In addition, the slack areas were formed along stratification and joints in the dip slope and destabilize the surface layer. Thus, the following two factors, (1) the penetration of igneous rock bodies and igneous dikes and (2) the dip slope formation, were main characteristics of the collapses of the Kumano Group [10].

Kuju Area topographically shows an undercut slope of the river, and it is highly possible that the erosion at the end of the slope encouraged the slack area in the surface layer of the Kumano Group. On the other hand, a fault fracture zone and a hydrothermal alteration zone have been observed at the lower part of the slope at Kuchitakata. It was suggested that the fault fracture zone and hydrothermal alteration zone might cause small collapses of the lower slope and encourage the slack zone in the surface layer of the upper slope.

The Kumano Group is distributed near the boundary with the Kumano acid igneous rocks and the Omine granitic rocks and has developed joints. Because of the large amount of sediment generated, the thick colluvial soils were deposited here. Also, Kumano acid igneous rocks in the upper part of the Kumano Group showed high relative elevation, and the thick collapsed acid rocks deposited on the rim were distributed. Since the areas of the collapsed rock deposits form mild slopes, they are easily extracted from 1:25,000-scale topographical maps. It is considered that these mild slopes collapse easily and the debris flows occur. Besides, the boundary between the colluvial soil and the Kumano Group of bedrock shows a typical topography affected by the geological structure; it is likely to collapse at the boundaries during heavy rainfall. It was necessary to give attention to the catchment areas where colluvial soils were distributed on the negative landform [12].

Debris flows in Kuju Aeas occurred in the landslide area judging from the landslide topography distribution map made by the National Research Institute for Earth Science and Disaster Prevention. The landslide has also been confirmed in the landslide area in Kuchitakata. It is suggested that the landslides in the Kumano group often occurred with topographical and geological factors such as erosion of undercut slope, faults, fracture zones, and dip slope. In other words, it is effective to extract landslide topography in the Kumano Group area as having a high risk of landslide slopes. However, landslide topography was not clear, and the landslide topography of the Kuchitakata collapse area was difficult to extract from topographic maps with a scale of 1:25,000. Although it was confirmed as a small flat area in the aerial photographs, it is also necessary to confirm the check points by a field survey. It was not easy to extract these areas from the wide area map.

5.6 Damage to Structures around Rivers

5.6.1 Progress in River Improvement and Scale of Flood Damage

Figure 5.5 shows the number of inundated houses divided by the total rainfall amount for inundation damage caused by rainfall of more than 200 mm in Wakayama Prefecture at the Showa Kii Peninsula Disaster (July 1953) to the Heisei Kii Peninsula Disaster (September 2011). In this publication, it is assumed that the number of flooded houses divided by the maximum total amount of rainfall observed in Wakayama Prefecture is an index to judge the extent of damage to the intensity of heavy rainfall. Dams were constructed along the Arita River, Futa River, and Hidaka River after the Showa Kii Peninsula Disaster (1953) that caused extensive damage in the prefecture. The Futagawa Dam along the Arita River was constructed in 1967, Hirokawa Dam along the Hiro River in 1975, and Tsubakiyama Dam along the Hidaka River in 1988. Therefore, the ratio of inundated houses to the intensity of heavy rainfall has decreased as a result of dam construction, as shown in Figure 5.5.

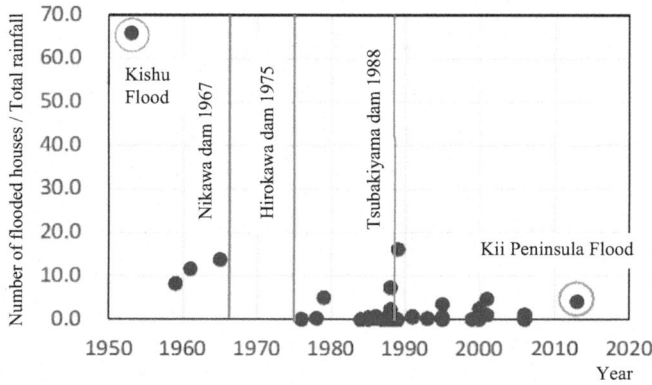

Figure 5.5 Scale of flood disasters in Wakayama Prefecture.

The maximum daily rainfall and total rainfall in Wakayama Prefecture at the Heisei Kii Peninsula Disaster in 2011 were almost twice as large as those of the recorded data of the Showa Kii Peninsula Disaster in 1953 as shown in Table 5.2 [13], but the number of flooded houses was only about 18%, which was less than that at the Showa Kii Peninsula Disaster. From this result, it is suggested that the improvement of flood control facilities contributed to reducing the damage even against the greatly exceeded ("unexpected") rainfall caused by S.T.S. Talas in 2011.

In the Hidaka River Basin, where was heavily damaged by the Kii Peninsula Flood, a cumulative rainfall of 999 mm has been recorded in the basin, this record exceeded the cumulative rainfall of 665 mm at the Kishu Flood in 1948. The amount of inflow and discharge into the dam lake almost was in equilibrium from midnight to 6:00 a.m. on September 4, therefore, the flood control capacity during this period was almost zero as shown in Figure 5.6 [14]. Thus, the river water level drastically rose at the lower area from the dam due to the exceeding flow over the estimated flow rate. The overflow, overtopping, and inundation occurred in the lower area.

Table 5.2 Comparison between Showa Kii Peninsula Disaster in 1953 and Heisei Kii Peninsula Disaster in 2011

Disaster Names	Maximum Daily Rainfall (mm)	Total Rainfall (mm)	Number of Houses under Water
Showa Kii Peninsula Disaster in 1953	564.7	665.6	43,804
Heisei Kii Peninsula Disaster in 2011	909	1,998	7,812

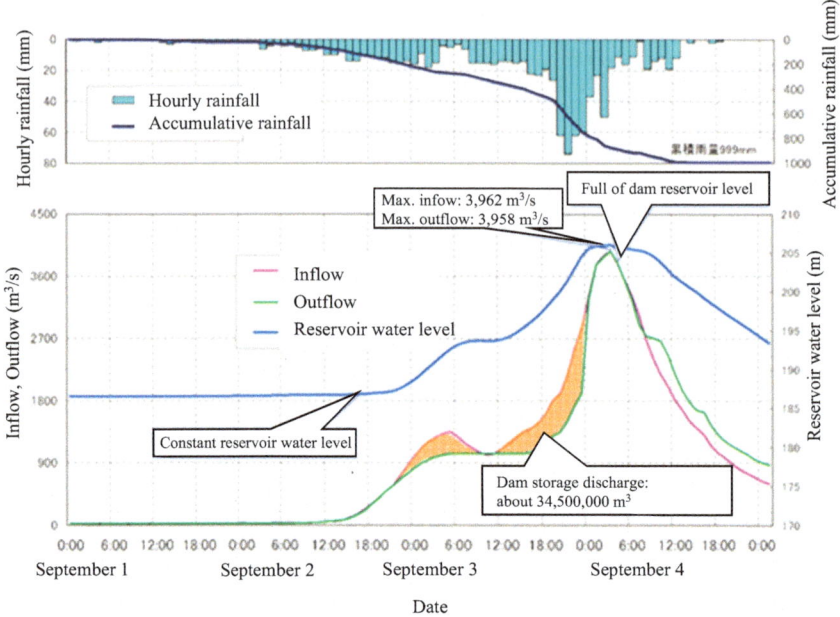

Figure 5.6 Flood control of Tsubakiyama dam in Hidaka River at the Kii Peninsula Flood.

5.6.2 Damages Caused by 'Unexpected' Heavy Rain

River structures are designed and constructed according to the estimated flow rate determined by probability theory. Although the river channels are stable under the estimated flow rate range, when the flow rate exceeds the estimated flow rate, the channels become unstable and damages such as riverbed scouring and overflow occur. Large-scale and severe damage was caused to the river basin due to the unexpected heavy rain. The damages of major flood disasters in the 2011 Kii Peninsula Disaster are described below.

5.6.3 Damage of the River Levee

At the Kirime River levee, the inner embankment covered by concrete blocks collapsed and slid inside at the middle part of the embankment (Photo 5.4). Because there was no trace of scouring at the failure site, the main reason for the embankment failure covered by concrete blocks was not overflow; the arc sliding failure occurred inside the embankment due to the rise of the water level inside the embankment and the decrease in the shear strength of the backfill soil from the high water level caused by the long-term heavy rainfall (Figure 5.7).

Photo 5.4 Failure of the block-protected embankment in Kirime River (Photo by Yasuyuki Nabeshima).

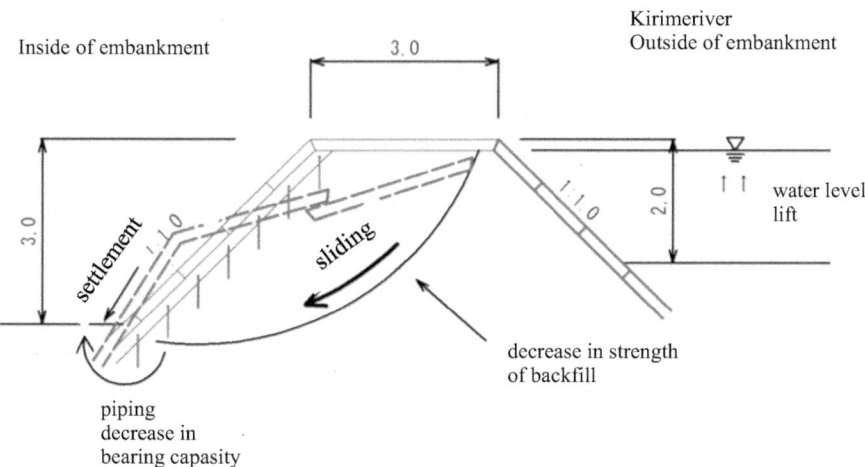

Figure 5.7 Estimation of damaged river embankment.

Figure 5.8 Collapse of EPS blocks along Kitayama River.

5.6.4 Collapse of EPS Backfill

The damaged point of National Route 169 was located about 500 m up-stream from the confluence of the Kitayama and Kumano Rivers. It was reclaimed in the recent road improvement work, and EPS (Styrofoam block) was installed as a backfill material in the retaining wall. EPS blocks were scattered, and the road was destroyed (Figure 5.8) by the Kii Peninsula flood. Judging from the sediment of fine sand on the road surface and the soil marks on the guardrail, it was suggested that the water level reached above the road surface during the flood. In this area, the water level of the Kitayama River increased above the level of the road surface, resulting in the floating and collapse of the EPS embankment by the action of buoyancy and scavenging forces beyond what it was designed for.

5.6.5 Bridge Damage (Pier Displacement, Overturning)

The center part of the riverbed ground at the piers of the old Hiki Bridge (Hiki, Shirahama Town) was scoured and the piers were dropped (Photo 5.5). Although the riverbed was stable in the case of conventional floods, it was suggested that the riverbed became unstable due to the unexpected flow rate and excess flow velocity. Consequently, the surrounding ground of the pier foundations were scoured, and the piers of the bridge were dropped.

5.6.6 River Maintenance and Damage Caused by 'Unexpected' Heavy Rains

Figure 5.9 shows the topography of the damaged areas at the time of Severe Tropical Storm Talas in 2011 and at the same areas in 1976 [15]. The Wakayama

Photo 5.5 Damage of piers of the old Hiki-Ohashi bridge (Photo by Noriaki Nakanishi).

Figure 5.9 Topographical change in Kawahara Area in Hidakagawa Town.

prefectural road of Gobo Miyama line was newly constructed along the river side, and the Minase River Bridge was built in 1976. These facilities were designed according to the estimated high-water levels and external forces factoring in the adjustment capacity of Tsubakiyama Dam, which was constructed in 1988. Consequently, when the unexpected heavy rain exceeded the dam adjustment capacity, many roads, bridges, and houses were severely damaged.

5.6.7 'Unexpected' Heavy Rain and Damage to Structures around Rivers

Structures around rivers are designed and constructed on the assumptions of estimated flow rates and estimated high-water levels, which were based on financial reasons. However, estimated floods are determined based on the probability theory and the rainfall record on the 100-year scale, according to the recorded rainfall in the Kii Peninsula heavy flood.

The river inundation area in the Kii Peninsula heavy flood corresponds to the inundation trace and river trace estimated from the topography. From this fact, it is possible to estimate the flood area caused by "unexpected" heavy rain exceeding the estimated rainfall scale by topographical interpretation, etc., and to recognize it as "possible flood damage." In recent years, as the rainfall has become more severe, it is necessary to prepare a recovery plan for structures and residential areas around rivers on the assumption that floods will exceed the estimated scale. Also, it is considered that the embankment retaining wall along the Kitayama River, which was a part of the road, was a suggestion for the river structure design from a viewpoint of quick restoration after the disaster. The structure reduced the risk of the damages occurring and made for quick recovery of damages after the disaster.

5.7 Damage to Cultural Property

5.7.1 *Cultural Properties of Wakayama and the Damage due to Severe Tropical Storm Talas*

The cultural properties of Wakayama Prefecture consist of 80 buildings designated as national treasures or important cultural properties, 201 registered cultural properties (buildings) designated by the prefecture, and 48 nationally designated historical sites, places of scenic beauty, or natural monuments (as in November, 2015). Many of them are concentrated in the northern part of the prefecture, while in the middle to southern Hidaka, Nishimuro, and Higashimuro Areas, there are several structures such as Kumano Hongu Taisha Shrine and Kumano-Nachi Taisha Shrine, and natural monuments and historical sites, which is a characteristic distribution situation. More than 10 buildings, historical sites, and places of scenic beauty among the cultural properties in Wakayama Prefecture are also registered as component assets of the "Sacred Sites and Pilgrimage Routes in the Kii Mountain Range," a World Heritage site, along with the Kumano Pilgrimage Routes.

Severe Tropical Storm Talas caused damage such as falling trees and collapses in Koyasan-Choishi in the northern part of the prefecture, flooding of Oyunohara and other places in Kumano Hongu Taisha Shrine in the southern part of the prefecture, burial of shrines in Kumano Nachi Taisha Shrine, sediment collapse and funeral hall collapse in Nachi Otaki Falls, disappearance of Mongaku Falls, debris flows in the Nachi Primitive Forest, and collapses in various parts of Kumano Pilgrimage Road, as well as broken trees and broken railings in Esuzaki and Shingu as shown in Figure 5.10.

5.7.2 *Damage to Kumano Nachi Taisha Shrine*

Slope failures, debris flow, and fallen trees around Kumano Nachi Taisha Shrine, a World Heritage Site, led to the burying of a shrine building as

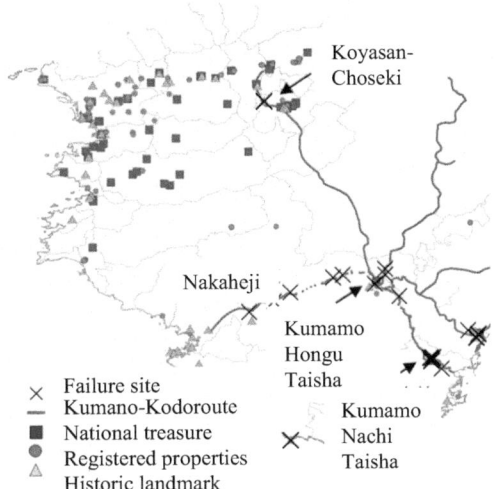

Figure 5.10 Cultural properties in Wakayama Prefecture and damages of properties.

Photo 5.6 Damage of shrine (the 5th building) and recovery of the slope (Photo provided by Kumano Nachi Taisha).

shown on the left side in Photo 5.6, the collapse of a water fence, the tilting of a gate, and the breakage of a box tower, as well as multiple shoulder collapses along the Koya Skyline Road leading to the shrine. The area around Kumano Nachi Taisha Shrine has landslide terrain, and several steep slope failure hazard areas and debris flow hazard mountain streams have been designated. In preparation for heavy rain disasters, a self-recording rain gauge has been installed at Kumano Nachi Taisha Shrine, and according to the records and interviews with duty personnel, it is considered that the heavy rain with hourly rainfall of 90 mm/h and cumulative rainfall of about 600 mm caused the debris flow and the collapse of the back of the shrine building due to the collapse (about 30 m wide, about 50 m long, about 15 m deep) that occurred near the summit of Mt. Myoho in the southwest around 2:30 p.m. to 3:00 p.m. on September 4, 2011. The geology of the origin of the debris flow is Kumano acid igneous rocks (granite porphyry) and various structures such as loose surface layers (strongly weathered and collapsed soils), shear planes, openings, viscous soils thought to be hydrothermally altered; and base rocks with different joint development states were identified in the slide cliffs. There were few catchments for rainwater, and it is thought that rainwater infiltrated deep into the ground from loose surface areas and developed joints, and that the rise in pore water pressure caused the collapse of a relatively large scale including the base rock. The stream path where the debris flow occurred in this disaster was continuous to the precinct in the past, but it currently changes its flow path to the outside of the precinct, and it returned to the continuous path to the precinct when the disaster occurred; therefore, it was difficult to predict the possibility of the water flow reaching the precinct.

The restoration work was carried out mainly by priests and students of equivalent Shinto priest training schools, since there are sacred areas such as the shrine building. Masonry walls were built on the back slope of the shrine building, and erosion control weirs were built at areas invisible from the shrine along the mountain steam. The damaged structures were also restored to their original situations as shown on the right side in Photo 5.6.

5.7.3 Future Tasks for Disaster Prevention of Cultural Heritages

The Sendai Framework for Disaster Risk Reduction from 2015 to 2030 was adopted at the 3rd UN World Conference on Disaster Risk Reduction, which was held in March 2015. In this statement, it was urgent and important to systematically anticipate risks, reduce risks, and strengthen resilience in order to more effectively protect cultural heritages (in the original text, people and communities and so forth were included) [16]. Here are some technical and financial tasks: Technical tasks include improving the accuracy of the prediction of sediment disaster points and occurrence conditions (such as rainfall), and information disclosure. Disaster prevention for cultural heritages should be carried out through cooperation between the owners and managers of cultural properties and experts in disaster prevention and by having a disaster-prevention plan and measures integrated with surrounding areas, which depend on the situation. However, in case there is no cooperation, the disaster prevention at individual cultural heritage sites becomes possible by preparing information sources that can, first and foremost, spot risks without expertise. Although there are many tasks for the active disaster prevention of cultural heritages, such as modification, landscape, and economic aspects, because of the variation of styles and managers of cultural heritage, it is necessary to promote prevention from the perspective of what can be done, such as at Kumano Nachi Taisha Shrine, where the rainfall was self-measured to help staff and tourists evacuate.

5.8 Summary

In this chapter, outlines of sediment disasters, such as large-scale slope failures, surface failures, debris flow, and river disasters, and cultural heritage damages were described. The large-scale slope failure occurred under conditions as described in 5.4, which are prone to large-scale slope failure, and it is necessary to pay close attention to valley-shaped slopes and concave slopes in the vicinity of thrust and fault in predicting future collapse. In addition, a new large-scale slope failure occurred with an accumulated rainfall of about 800 mm, depending on geological conditions, and this result can be referred to in the construction of a warning and evacuation system. The surface failures and debris flows were characterized by geologic groups such as Kumano acid igneous rocks and the Kumano formation group as described in Section 5.5. Many collapses in the Higashimuro Area are concentrated in the eastern area of the Kumano acid igneous rocks. In contrast, fewer sediment disasters occurred in the Kumano formation group than with the Kumano acid igneous rocks, and the scale of collapses was also smaller. Regarding the river disasters, the number of flooded houses at the Heisei Kii Peninsula Disaster in 2011 was less than that at the Showa Kii Peninsula Disaster in 1953 because the improvement of flood control facilities contributed to reducing the damage. However, from the damage of structures around rivers, it is necessary to prepare a recovery plan for structures and residential areas around rivers on the assumption that floods will exceed the estimated scale. Also, some of the cultural properties in Wakayama Prefecture, such as Kumano Nachi Taisha Shrine, were damaged due to sediment disasters. Disaster prevention for cultural heritages requires cooperation between the owners, the managers of cultural properties, and the experts, and the disaster-prevention plan and the measures need to be integrated with surrounding slopes.

References

[1] Yoshimatsu, T., Nakaya, S., Kodama, T., Terai, K. and Harada, T.: Geology and Hot Springs in the Kii Peninsula, *Urban Kubota*, Vol. 38, p. 56, 1999. (in Japanese)
[2] Research Group of Kishu Shimanto: New Developments of Shimanto Adducts in the Kii Peninsula, *Jidanken Senpo*, Vol. 59, p.310, 2012. (in Japanese)
[3] Suzuki, R., Egusa, N., Yano, H., Tsujino, H., Nabeshima, Y. and Ushiro, S.: Relationship between sediment disaster and rainfall in Wakayama Prefecture due to Typhoon No. 12 in 2011, *Kansai Geo-Symposium 2013*, pp.87–90, 2013. (in Japanese)
[4] Yano, H., Nabeshima, Y., Nonami, S., Tanigaki, K., Ishida, Y. and Ushiro, S.: Characteristics of the large scale slope failure in the Hidaka-Nishimuro Area of Wakayama Prefecture caused by Typhoon No. 12 in 2011, Annual conference of JGS, pp.2255–2256, 2015. (in Japanese)

[5] Suzuki, H., Harada, T., Isonokami, T., Kumon, F., Nakatani, S., Sakamoto, T., Tateishi, M., Tokuoka, T. and Iuchi, Y.: Geological survey of the Kurisugawa area, 1/50,000, Geological Survey, 1979. (in Japanese)

[6] Kinoshita, A., Kitagawa, S., Oyama, M., Kosugi, K., Uchida, T., Kanemura, K., Okajima, S. and Okude, T.: A study on the hydrological characteristics of a collapsed slope in the Kumano area, where a deep collapse occurred during Typhoon No. 12 in 2011, Proceedings of the Japanese Society of Erosion Control in 2013, pp. A-76–A-77, 2013. (in Japanese)

[7] Sakurai, W., Sakai, R., Iwata, K., Kosugi, K., Kanamura, K. and Okajima, S.: Relationship between deep collapse and geological and hydrological properties in Kumano, Proceedings of the Japanese Society of Erosion Control in 2014, pp. A-28–A-29, 2014. (in Japanese)

[8] Nonami, S., Nabeshima, Y., Ushiro, S., Tanigaki, K., Yano, H. and Ishida, Y.: A case study of geological structure and collapse mechanism in the Fushuno district of Wakayama Prefecture, *Kansai Geo-Symposium 2014*, pp.189–194, 2014. (in Japanese)

[9] Egusa, N., Tsujino, H., Tanigaki, K., Izuminami, R., Yano, H., Kato, T., Iwase, N., Ishida, Y., Fujimoto, M., Naoto Iwasa, N. and Ushiro, S.: Debris flow disaster in the Higashimuro area of Wakayama Prefecture caused by Typhoon No.12 in 2011, *Kansai Geo-Symposium 2014*, pp.179–184, 2014. (in Japanese)

[10] Egusa, N.: Landslides in Wakayama Prefecture Caused by Typhoon No.12 in 2011, *Soil and Rock*, Vol. 63, pp.26–34, 2015. (in Japanese)

[11] Fujii, S., Egusa, N., Ishida, Y. and Ushiro, S.: Characteristics of slope failures in the Higashimuro area of Wakayama Prefecture caused by Typhoon No.12 in 2011, Annual conference of JGS, pp. 2277–2278, 2015. (in Japanese)

[12] Tsujino, H., Egusa, N., Ushiro, S., Tanigaki, K., Iwase, N.: On the actual condition and topographical features of collapse and debris flow in the Kumano Group distribution area in the Higashimuro Area of Wakayama Prefecture, *Kansai Geo-Symposium 2015*, pp.147–150, 2015. (in Japanese)

[13] Wakayama Prefecture Information Center: Disaster History of Wakayama Prefecture (after the Meiji period), <http://www.pref.wakayama.lg.jp/prefg/080100/saigai/saigai01.htm.> (referenced 2015/12/01)

[14] Wakayama Prefectural Information Center: About Flood Control at Futagawa Dam, Hirokawa Dam, *Tsubakiyama Dam and Nanagawa Dam during Flood Flooding Associated with Typhoon No. 12*, <http://www.pref.wakayama.lg.jp/prefg/080400/kouzuichousetu/H23/taihuu12.html.> (referenced 2015/02/28)

[15] Geographical Survey Institute: Information Aerial Photographs on the Damage Caused by Typhoon 12, <http://zgate.gsi.go.jp/saigaishuyaku/20110906/index2.htm.> (referenced 2015/11/08)

[16] The Ministry of Foreign Affairs: Sendai Framework for Disaster Risk Reduction 2015–2030, p.4, <https://www.mofa.go.jp/mofaj/files/000071589.pdf.> (referenced 2023/03/17)

6 Disasters in Mie Prefecture

Kenji Okajima

6.1 Introductions

Severe Tropical Storm Talas (1112), which struck from the end of August to the beginning of September 2011, resulted in 2 deaths, 1 missing person, 55 houses being completely destroyed, 1,683 houses being flooded above the floor level, and 972 public works facilities being damaged in Mie Prefecture [1]. This torrential rain caused large-scale slope failures in Matsusaka City, Odai Town in Taki County, and Kihoku Town in Kita Muro County. Furthermore, in the Higashi-Kishu Region, ranging from Owase City to Kiho Town, many surface failures and accompanying debris flows were observed, along with the flood damage due to overflowing rivers.

In this chapter, the following sediment disasters that were caused by the S.T.S. Talas in Mie Prefecture are discussed: Large-scale slope failures in Higashimata Valley located in Odai Town, Taki County and Kajiyamata Valley located in Kihoku Town, Kita Muro County; a number of surface failures in the Higashi-Kishu Region stretching from Owase City to Kiho Town; and damages to the Kumano Kodō region caused by a large-scale slope failure at the Yokogaki Pass in Mihama Town. Further, this chapter describes the following river disasters: A failure of a ring levee of the Onodani River and failures and overflow of levees at the Omata River and Ido River in Kumano City and Kumano River in Kiho Town.

6.2 Large-Scale Slope Failures

6.2.1 Large-Scale Slope Failure at Higashimata Valley

Photo 6.1 shows the large-scale slope failure at the Higashimata Valley in Odai. Higashimata Valley is located in an area where fractured alternations of sandstone and mudstone containing chert and mélange of the Chichibu zone accretionary complex are present. Figure 6.1 shows a comparison between a site condition map based on field survey results and a three-dimensional diagram prepared from aerial laser survey results (hereafter referred to as LP diagram).

Figure 6.2 shows a cross-sectional view in the direction of the main survey. The elevation at the top of the failed area was 750–760 m. The scale of the slope

DOI: 10.1201/9781003375210-6

Photo 6.1 Large-scale slope failures at Higashimata Valley (Photo by Masaki Ishikawa).

failure was quite large: 700 m long (horizontal distance), 350 m high, 300 m wide, and 50 m thick. It includes the unstable areas on the back of the failed area and the slide cliff above the top, as confirmed by the field survey and LP diagram. The field survey confirmed the existence of bedding plane faults (thrusts) at the bottom of the failed area with an apparent inclination of 20°–30° to the slope, and the development of the bedding planes and joint planes creates flow surfaces with a slope of 35° to 50° in the sandstone-dominant sandstone-mudstone alternation above the slope. The pre-failure topography around the failed site showed signs of past deformations, such as a gentle slope at the top and a double ridge, suggesting that bedrock creep was in progress.

In addition, an old failure landform was observed at the northwestern end of the slope that existed prior to the current one, and it is possible that the failure was expanded due to the erosion by streams, causing instability. At an elevation of approximately 415 m on the right bank and approximately 440 m on the left bank, along the stream below the failed site, there were alternating layers of unconsolidated mud, sand, and gravel mixed with cobblestones, which are assumed to be aquatic deposits. Their characteristics indicate that they are debris flow deposits left in a flood plain formed by river channel blockage in the past and flood plain deposits made under static conditions. Carbon-14 (14C) dating of the carbonized wood incorporated within the alternating layers indicates that the wood was deposited during the last glacial

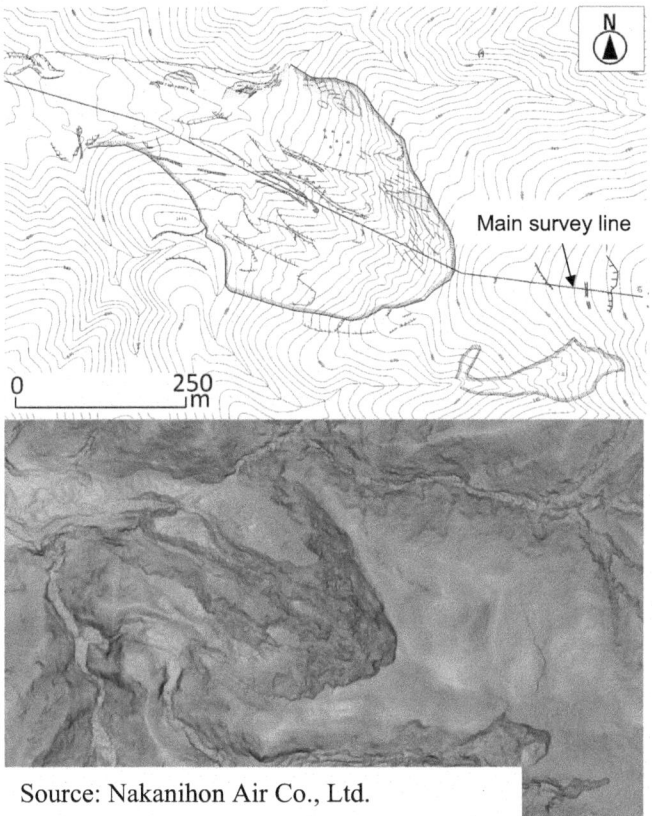

Source: Nakanihon Air Co., Ltd.

Figure 6.1 Plan view and LP diagram of Higashimata Valley.

period, dated between 18,895 and 18,623 years ago. The 14C dating of the wood chips collected from the upstream area of this location indicated that it was 200 to 300 years old, suggesting that there are several layers of sediment in the flooded area and that sediment transport has occurred repeatedly at this location. Furthermore, "slope failure landforms" can be seen on both banks of the valley in the downstream area, as shown in Figure 6.3, and a flat surface with a similar elevation, which is unnatural in this topography, can be seen near the elevation indicated by the black circle in the figure. Thus, it is surmised that a natural dam may have formed near the slope failure in the past, and that a large lake may have existed in the upstream area of the dam because of blockage of the river channel.

6.2.2 Conditions of the Kajiyamata Valley

Photo 6.2 shows the large-scale slope failure at the Kajiyamata Valley, Kii-nagashima Area, Kihoku Town. Kajiyamata Valley is located in the central

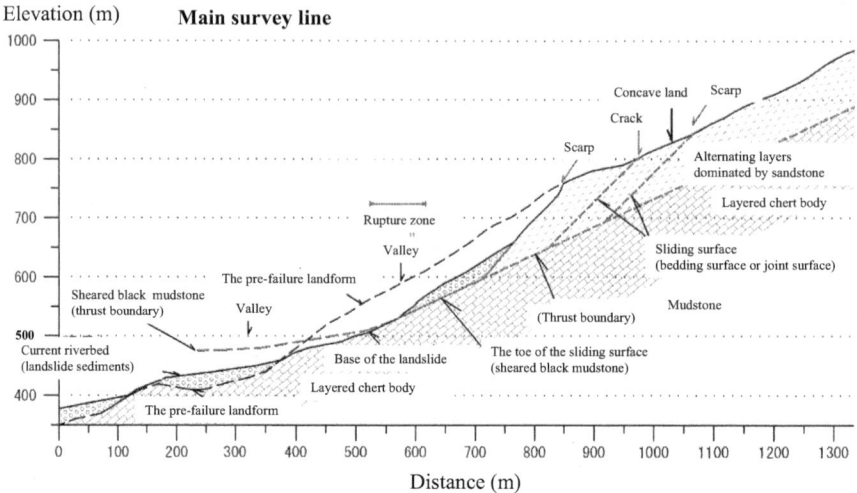

Figure 6.2 Cross section of Higashimata Valley.

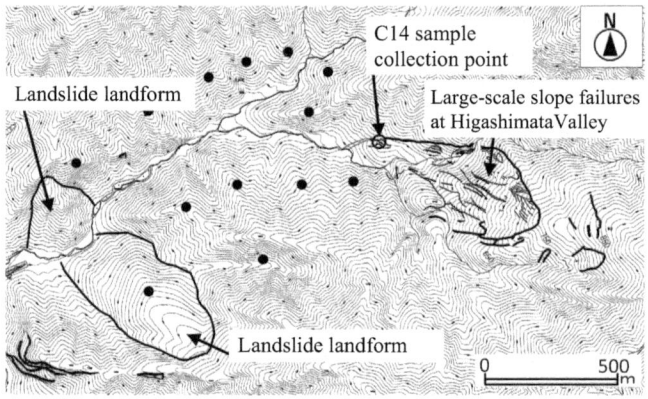

Figure 6.3 Conditions around the area of large-scale slope failure of Higashimata Valley.

unit of Matoya Group of the Shimanto Belt accretionary complex, and is mainly composed of stratified or fractured sandstone-mudstone alternations and sandstones, with muddy mélange. Many thrusts are formed in the muddy mélange. Figure 6.4 shows a comparison between the site condition maps based on the field survey results and the LP diagrams prepared from the aerial laser survey results. Moreover, Figure 6.5 shows the cross-sectional view in the direction of the main survey. The elevation of the top of the failed site was about 620 m. The large-scale slope failure is considered to have occurred on the upper slope from an elevation of around 410 m halfway up

Photo 6.2 Large-scale slope failure at Kajiyamata Valley (Photo by Masaki Ishikawa).

the slope, with a length (horizontal distance) of about 420 m, a height of about 220 m, a width of about 150 m, and a thickness of about 30 m. The geological structure of the east side of the failed site encompassing Tochiko Valley, confirmed by the fieldwork, shows a bent structure around the failed site, as shown in Figure 6.6, which was consistent with the topography. The strike of the bedding plane of strata identified in the failed area and the fault were inclined from northeast to the southwest direction at 35°–45°, thereby presenting a flow-bed structure parallel to the slope. The LP map of this location shows a double ridge on the top west side of the failed site, suggesting that bedrock creep had been ongoing here prior to the slope failure.

6.2.3 Rainfall Conditions in the Vicinity of the Large-Scale Slope Failures

In September 2004, Typhoon Meari (0421) caused numerous slope failures in the former Miyagawa Village, which was located in the area where the current large-scale slope failure occurred. Figure 6.7 shows high precipitation volume at the Miyagawa Observation Station near the collapsed site during Typhoon Meari of 2004 and Severe Tropical Storm Talas in 2011. In comparison, the cumulative precipitation of Typhoon Meari in 2004 was approximately 800 mm, while that of Talas in 2011 was 1,600 mm. The precipitation data of Talas in 2011 shows that the rainfall of about 30 mm/h lasted over a four-day

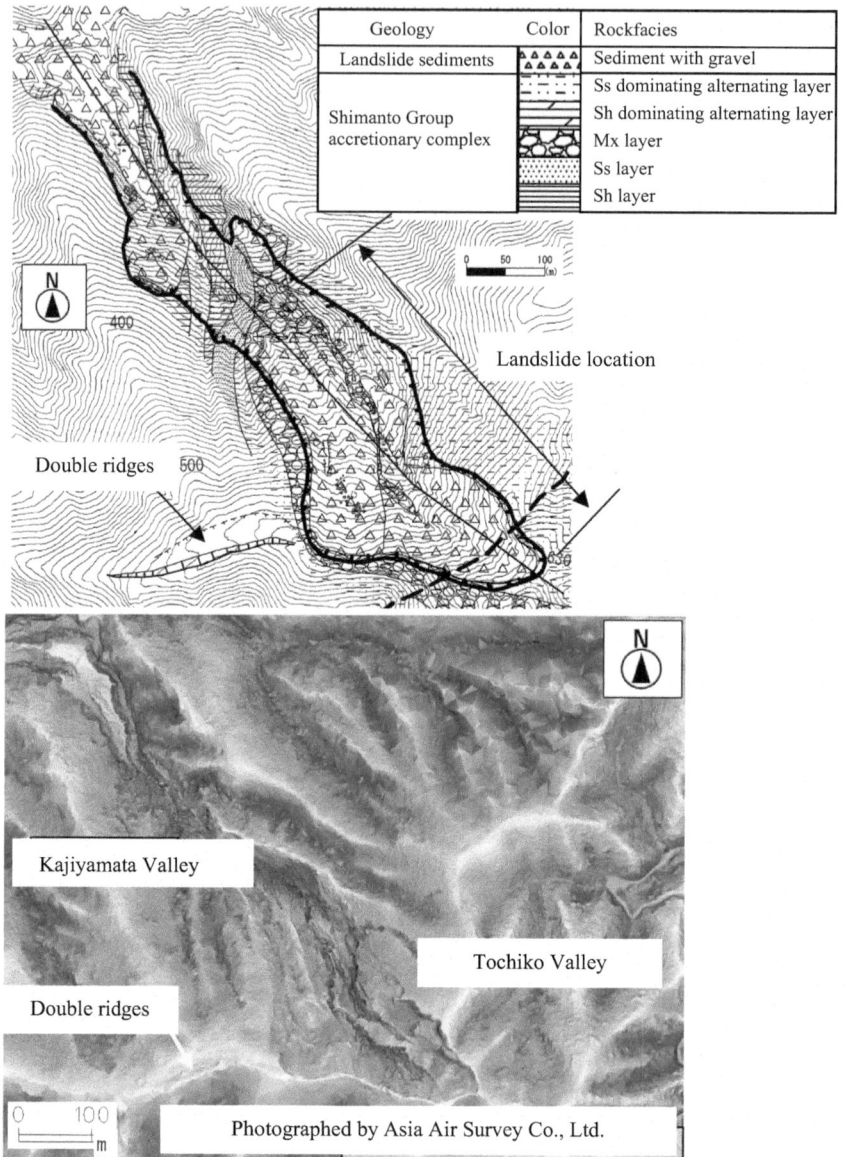

Figure 6.4 Plan view and LP diagram of Kajiyamata Valley.

period with a maximum hourly rainfall of about 80 mm/h. In contrast, the cumulative precipitation of Typhoon Meari in 2004 was about 800 mm, while the maximum hourly precipitation was as high as 119 mm/h. Aizawa et al. [2]. examined the relationship between the cumulative precipitation and the hourly

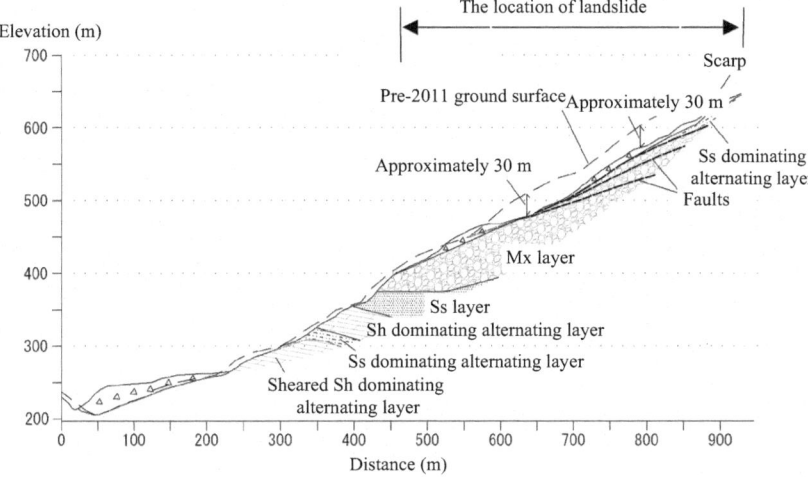

Figure 6.5 Cross section of Kajiyamata Valley.

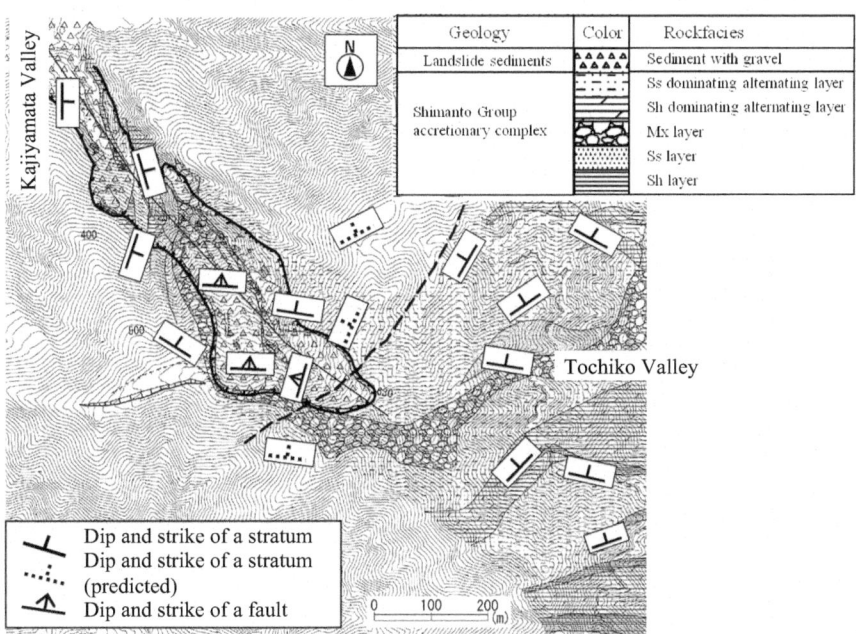

Figure 6.6 Conditions around the large-scale slope failure of Kajiyamata Valley.

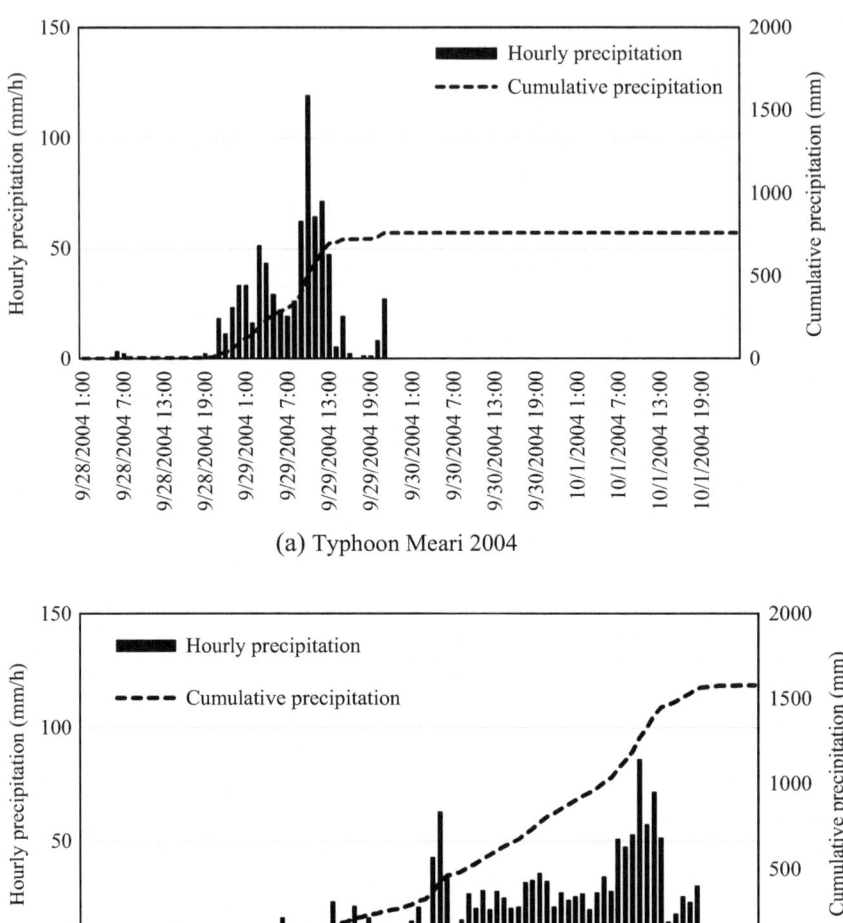

(a) Typhoon Meari 2004

(b) Severe Tropical Storm Talas 2011

Figure 6.7 Precipitations at the Miyagawa Observation Station.

precipitation rate with regard to the occurrence of slope failure during Typhoon Meari in 2004. They reported that although there was no clear relationship between the locations of the slope failures and the cumulative precipitation, many slope failures occurred in areas where the maximum hourly precipitation was 110–120 mm/h, and that the occurrence of slope failures was

Photo 6.3 Spring water near the thrust boundary [Higashimata Valley] (Photo by Masaki Ishikawa).

not so remarkable when the maximum hourly precipitation was about 80 mm/h. Meanwhile, differences in the failure morphology were observed; the slope failures due to Typhoon Meari in 2004 occurred mostly near the surface, whereas those due to S.T.S. Talas in 2011 were large-scale failures that extended to the deeper layers. During the field surveys conducted at Higashimata Valley and Kajiyamata Valley where large-scale slope failures occurred during Talas in 2011, the spring water shown in Photo 6.3 was found to flow out near the thrust boundary, which was considered to be the failed plane, suggesting the existence of water channels in the bedrock. In addition, the electrical conductivity of the spring water observed at the failed plane was measured to be as low as 4–5 mS/m, suggesting that the spring water might be rainwater discharging quickly through the water route within the bedrock. From these facts, it is presumed that rainwater due to continuous rainfall over a long period of time increases the groundwater level, leading to large-scale slope failures.

6.3 Surface Failures

6.3.1 Conditions of Surface Failures in the Higashi-Kishu Region

In the Higashi-Kishu Region in Mie Prefecture, many surface failures occurred in the area from Owase City to Kiho Town. The collapsed areas were

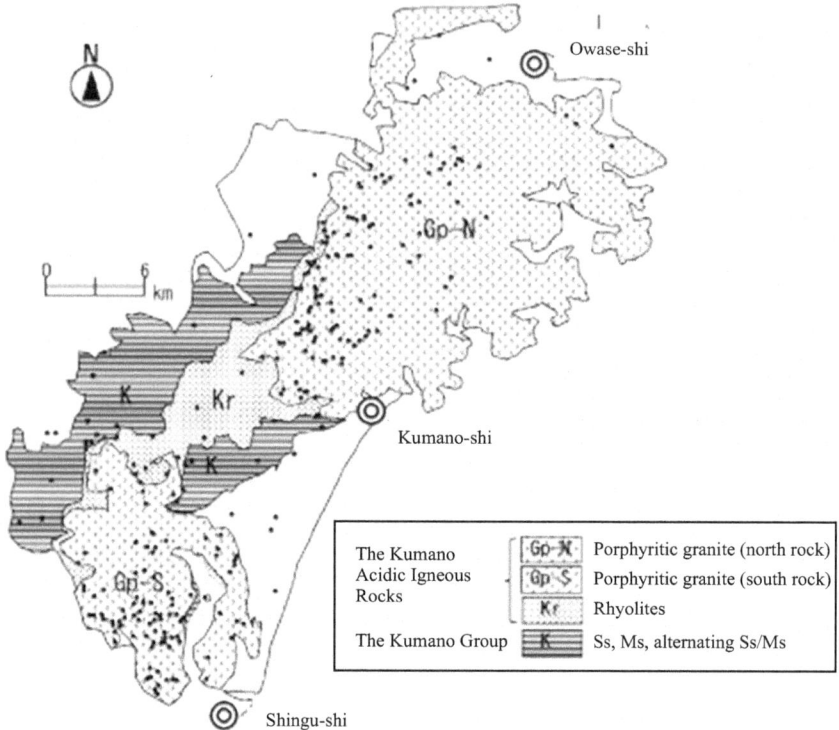

Figure 6.8 Geology and locations of slope failures.

examined by comparing the aerial photographs taken before (2006) and after (2011) the disaster based on a digital map of Mie Prefecture. The results are shown on a seamless geological map [3] (Figure 6.8). The geology of the area from Owase City to Kiho Town consists mainly of granite porphyry and rhyolite of the Kumano acidic igneous rocks, which are Neogene volcanic and plutonic complexes, and the Kumano Group, which is a tertiary sedimentary formation. Furthermore, granite porphyry of the Kumano acidic igneous rocks is divided into the northern formation around Owase City and Kumano City and the southern formation around Kiho Town [4]. When the slope failures were counted by rock type, a large number of slope failures in the areas consisting of granite porphyry of the Kumano acidic igneous rocks were discovered. In contrast, slope failures were extremely rare in the areas comprising rhyolites and the Kumano Group. Table 6.1 shows the number of slope failures by rock type and per 1 km^2. Ninety percent of all slope failures occurred in the areas consisting of Kumano acidic igneous rocks. Moreover, 80% or more of the slope failures among them occurred in areas consisting of granite porphyry. The density of slope failures per 1 km^2 was about 0.1 locations for the Kumano Group, followed by 0.4 locations for rhyolites

Table 6.1 Number of slope failures by rock types

Geology			No. of Slope Failures		Ratio (%)		Slope Failures / km²
The Kumano	Porphyritic	North rock	118	262	40.7	90.3	0.41
acidic igneous	granite	South rock	117		40.3		1.06
rocks	Rhyolites		27		9.3		0.44
The Kumano Group, Shimanto Group			28		9.7		0.10

and the northern formation of granite porphyry of the Kumano acidic igneous rocks. In contrast, the density was 1.06 locations for the southern formation of granite porphyry, showing the tendency of high frequency of occurrence.

The pre-failure slope gradient around the top of the failure was examined based on the digital elevation model (DEM) data of the digital map of Mie Prefecture before the disaster (2006) and presented in Figure 6.9. In the area consisting of the Kumano Group and rhyolites and the southern formation of granite porphyry of the Kumano acidic igneous rocks, the number of failures was higher on the slopes with the pre-failure gradient of 30°–40°. Meanwhile, in the areas consisting of the northern formation of granite porphyry, the number of failures was higher on the slope that was 5°–10° steeper than the former (i.e., the slope with the pre-failure gradient of 35° to 45°). Figure 6.10 shows the direction of slopes at the slope failure locations by rock types. Slope failures are concentrated on the south-facing slopes from southeast to southwest, accounting for 61% for the Kumano Group, 74% for rhyolites of the Kumano acidic igneous rocks, 87% for the southern formation of granite porphyry, and 80% for the northern formation.

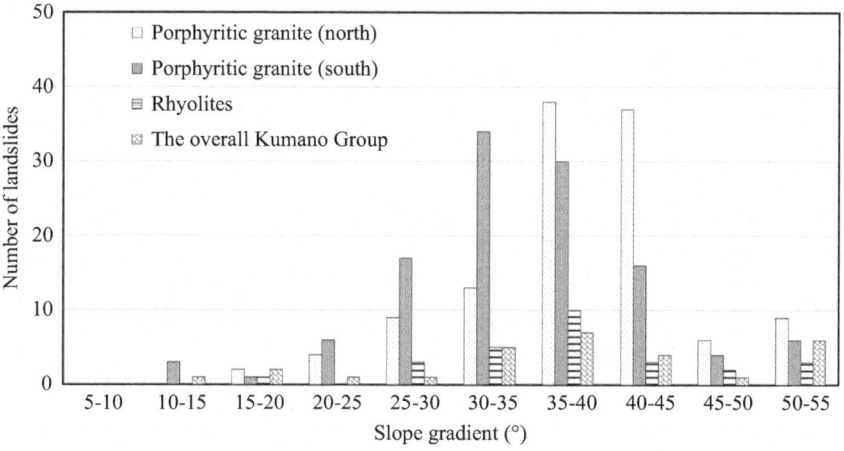

Figure 6.9 Slope gradient at the head of slope failure by rock types.

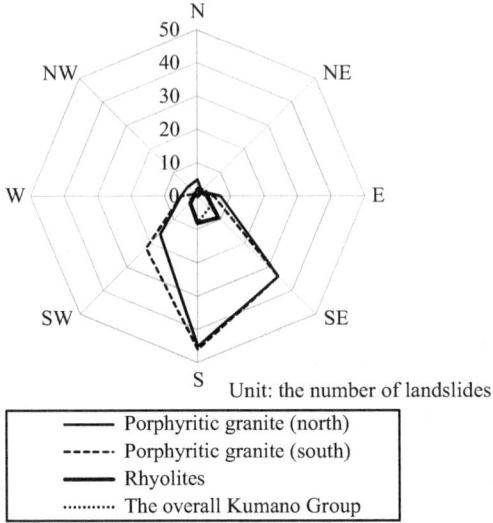

Figure 6.10 The direction of failure slope by rock types.

The horizontal cross-sectional profiles of the slopes were classified into a valley slope, a ridge slope, and a straight slope according to Suzuki [5] and summarized in Table 6.2. The results showed that among all rock types, the numbers of slope failures were higher on the valley slopes and the straight slopes, while those on the ridge slopes were low, accounting for only 5%. Field surveys were conducted in the area comprising of granite porphyry of the Kumano acidic igneous rocks, where the number of the slope failures were the highest, to confirm the geological conditions, nodal structure, gradient, and scale of the failures. Although a relatively large failure with a maximum width of 60 m and a maximum depth of 15 m was observed on the southern formation, the scale of the failures was relatively small in general, as shown in Photo 6.4. The average width was found to be 20 m or less, and the average depth was 5 m or less.

Table 6.2 Horizontal cross section of failure slopes by rock type

Geology			Valley Slope	Ridge Slope	Straight Slope
The Kumano acidic	Porphyritic	North rock	62	0	56
igneous rocks	granite	South rock	52	8	57
	Rhyolites		13	3	11
The Kumano Group, Shimanto Group			8	3	17
Total	Number of slope failures		135	14	141
	Ratio (%)		46.6	4.8	48.6

Photo 6.4 Typical surface slope failure [northern formation] (Photo by Yasushi Kataoka).

The location of the slip surface was considered to be near the boundary between talus deposits and bedrock or between heavily and moderately weathered bedrocks. In the northern formation, talus deposits and topsoil made up 75% of the failed geology, whereas in the southern formation, 57% was heavily weathered bedrock. This suggests that weathering may have progressed deeper in the southern formation than in the northern formation.

The inclination of strike of the joint surfaces in the northern formation was mainly centered in the south and toward the southwest to the southeast, while in the southern body, the strike tended to incline east-southeast to southeast and west-southwest to southwest, which is consistent with the direction of the slope at the slope failure locations, based on the field survey results.

On the site, cavities that appeared to be traces of piping with a diameter ranging from several centimeters to a maximum of 1 m were observed on the scarp at the top of the collapsed area, as shown in Photo 6.5. This indicates that a water channel had been formed on the slope at the failure locations. The electrical conductivity of the surrounding spring water and stream water was measured, and the results showed low values of about 3–4 mS/m in both the northern and southern rock formations. This suggests that the rainwater that passed through the ground may have run off directly as a spring without spending much time in the ground.

Photo 6.5 Piping seen on the main scarp (Photo by Yasushi Kataoka).

6.3.2 Relationship between Precipitation and Slope Failures at the Higashi-Kishu Region

Figure 6.11 shows the precipitation at the Osako Observation Station in Kumano City, which is located in the area consisting of granite porphyry of the Kumano acidic igneous rocks where multiple slope failures occurred,

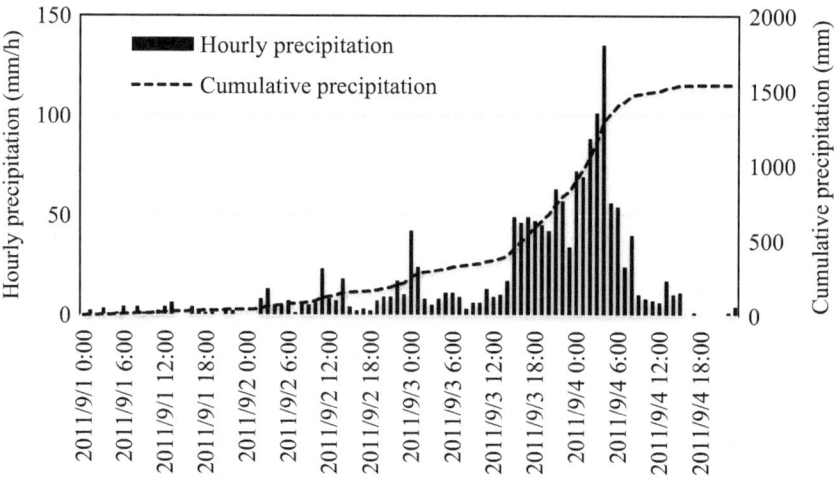

Figure 6.11 Changes in precipitation at the Osako Observation Station.

1–4 September, 2011. It shows a late onset of concentrated rainfall patterns with a rapid increase in precipitation starting on the night of September 3 and a maximum hourly precipitation of 135 mm/h from 3:00 a.m. to 5:00 a.m. on September 4. At 3:10 a.m. on September 4, when the maximum hourly precipitation was recorded, NTT's communication service reported that "landlines and Internet services were unavailable in the entire Kumano City, Mihama Town, and Kiho Town." Interviews with various parties in the area around Kiho Town also revealed that the reports of disaster were concentrated around the timeframe when the maximum hourly precipitation was observed.

In this connection, the relationship between the location of the slope failures and the amount of precipitation from 3 to 5 a.m. on September 4 was examined when the maximum hourly precipitation was recorded and disaster reports were concentrated. Figures 6.12 and 6.13 show isohyet diagrams of the hourly precipitation and the cumulative precipitation from 3–4 a.m. and 4–5 a.m., respectively, superimposed on the geological features and locations of slope failures. These figures show that the number of slope failures tended to be higher in the area consisting of granite porphyry of the Kumano acidic

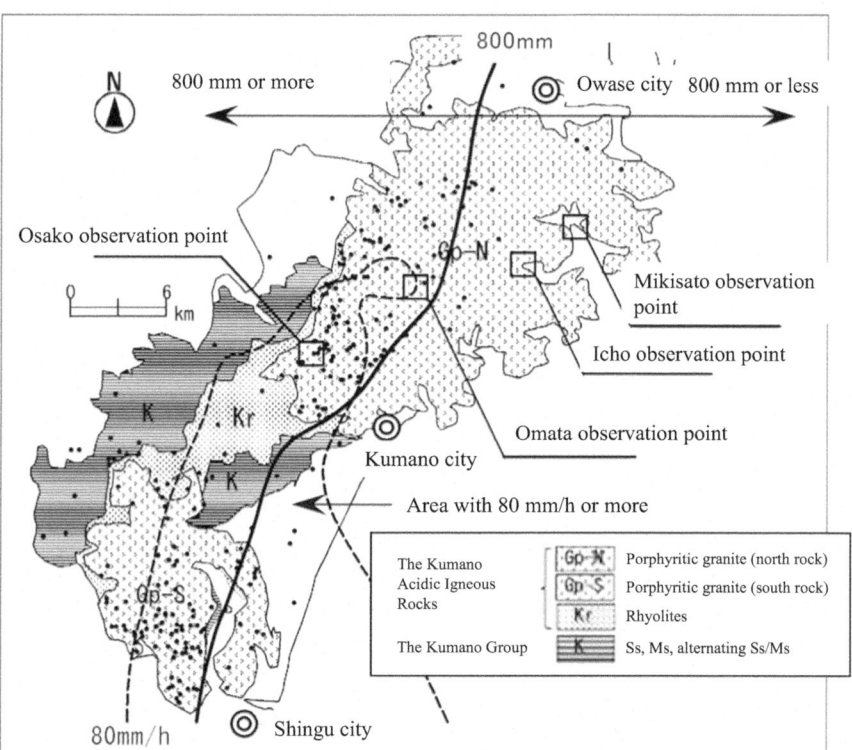

Figure 6.12 Isohyet diagram (1) [9/4, 3:00–4:00 a.m.].

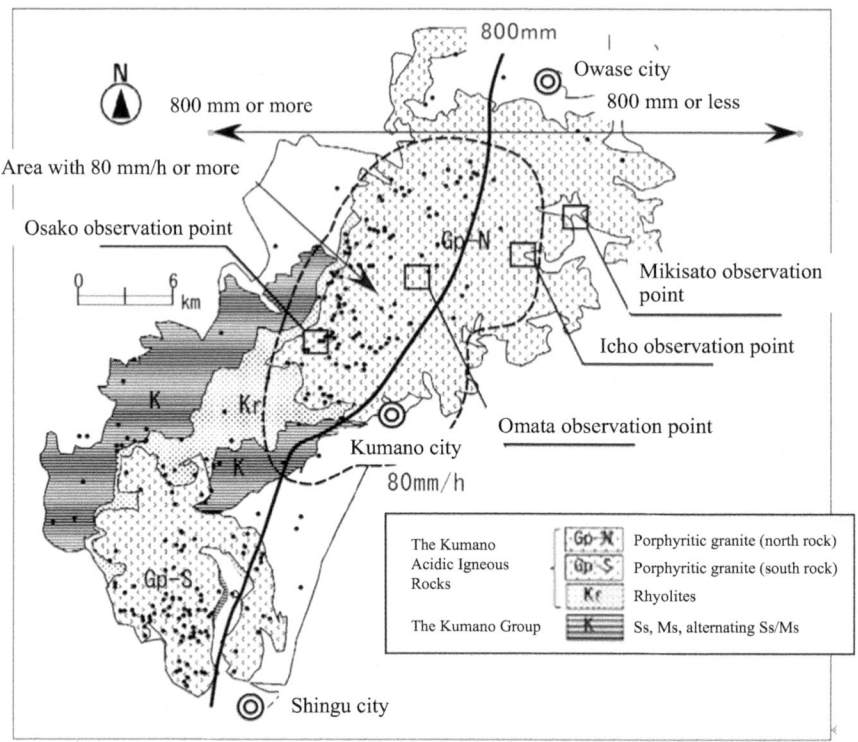

Figure 6.13 Isohyet diagram (2) [9/4, 4:00–5:00 a.m.].

igneous rocks, where the cumulative precipitation was about 800 mm and the hourly precipitation exceeded 80 mm/h.

Mikisato and Icho Observation Stations also recorded precipitation exceeding the hourly precipitation of 80 mm/h and the cumulative precipitation of 800 mm; however, there was hardly any slope failure in the surrounding area. Figure 6.14 shows changes in the cumulative and hourly

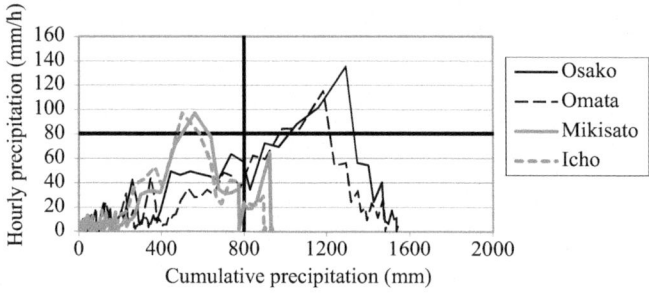

Figure 6.14 Changes in cumulative and hourly precipitation at each observation points.

precipitations between September 1 and 4 recorded at the Osako and Omata Observation Stations, where many slope failures were observed, and the Mikisato and Icho Observation Stations, where few slope failures were observed. The Osako and Omata Observation Stations recorded a cumulative precipitation exceeding 800 mm with an hourly precipitation of 80 mm/h. In contrast, at the Mikisato and Icho Observation Stations, the cumulative precipitation at 5 a.m. and 6 a.m. on September 4 was 502 mm and 564 mm, respectively, when the maximum hourly precipitation of 97 mm/h was observed, and the cumulative precipitation did not reach 800 mm during the time when rainfall exceeded 80 mm/h per hour. These results indicate that in the area consisting of granite porphyry of the Kumano acidic igneous rocks in the Higashi-Kishu Region, the late onset of concentrated rainfall exceeding 80 mm/hour, combined with a cumulative precipitation of 800 mm, may have caused multiple occurrences of the surface failures.

6.4 World Heritage

6.4.1 Yokogaki Pass on the Iseji Route, Kumano Kodō

The Kumano Kodō was registered as a world cultural heritage site on July 7, 2004, as part of the "Sacred Sites and Pilgrimage Routes in the Kii Mountains." The registered area covers 29 cities and towns in three prefectures, Mie, Nara, and Wakayama, and covers an area of 495.3 ha [6]. Yokogaki Pass is located on the Iseji route, which runs along the eastern side of the Kii Peninsula. It is an attractive route, with beautiful cobblestone paths in the hilly terrain over the pass from Ise Jingu Shrine to Kumano Sanzan. It is thought that the cobblestones that characterize the Kumano Kodō were installed by the people at the time as a disaster-prevention measure to prevent road runoff due to heavy rains in this area, which is known to be one of the heaviest rainfall areas in Japan. At the Yokogaki Pass, S.T.S. Talas of 2011 caused the Kodō and adjacent forest roads to slide down due to slope failure and to be buried due to soil and sand runoff in various places.

6.4.2 Damage to the Yokogaki Pass

Slope failures in multiple locations due to torrential rains were triggered by Severe Tropical Storm Talas in 2011, as shown in Figure 6.15. The maximum 24-hour precipitation at the Kounogi Observation Station between 2 p.m. on September 3 to 2 p.m. on September 4 was 716 mm, and the maximum hourly precipitation was measured at 98 mm/h from 4 a.m. to 5 a.m. on September 4. A relatively large-scale slope failure (collapse width of 95 m and Kodō runoff length of 150 m) occurred at location No. 1 on the Sakamoto side (Figures 6.15 and 6.16). The Kodō and the adjacent forest roads were lost due to slides and buried due to soil and sand runoff. In addition, at two locations near the center of the section, No. 2 and No. 3, the

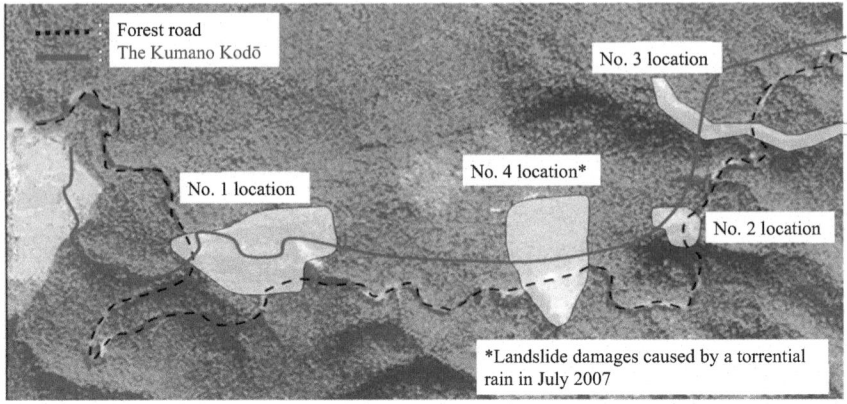

Figure 6.15 Yokogaki Pass and damage locations (Original map provided by Mie Prefecture; captions added by authors).

Figure 6.16 The head on the south side/scarp on the north side (No. 1 location) (Photo by Tatsuo Konegawa).

Kodō was lost due to slope failures and debris flows. The length of the Kodō got washed out in approximately 250 m of the Yokogaki pass, and the damage rate totaled 14%. Restoration of the collapsed and washed-out locations was practically impossible. Currently, the management group has set up detours wherever possible, and some of the roads are now open to traffic as an emergency measure.

6.5 River Disasters [7]

6.5.1 Omata River

Figure 6.17 shows an aerial view of the survey location. The area encompassed by the thick line on the left bank of the bridge in the photograph is

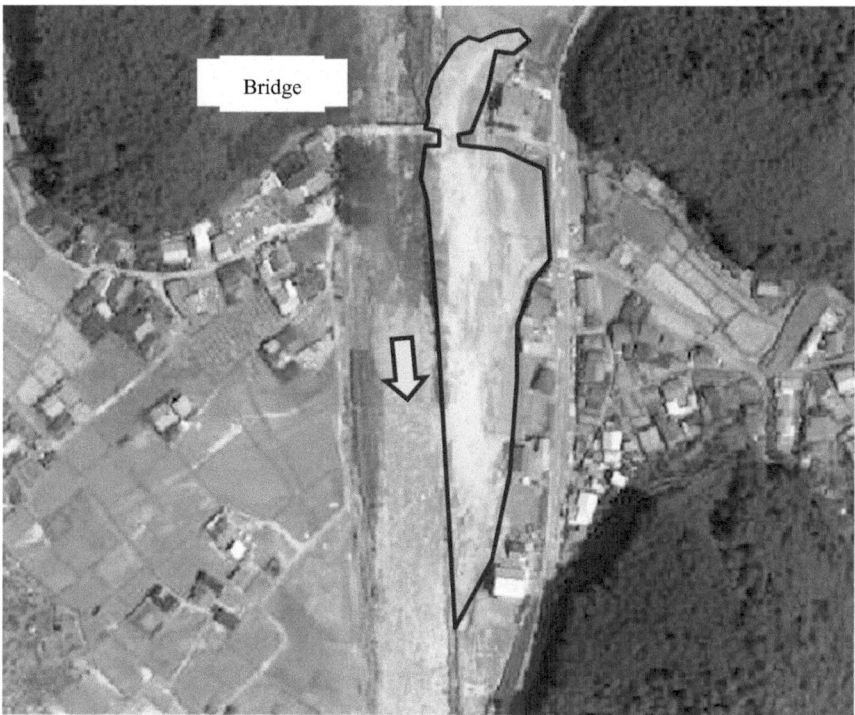

Bridge

Figure 6.17 The area around the survey point (Reprinted from an aerial photograph by Geospatial Information Authority of Japan, captions added by authors).

the inundated area. The Omata River disaster was considered to have occurred when the river channel was blocked by driftwood on the bridge piers and the stagnant flood waters overflowed the left bank levee upstream of the bridge. The overflowing floodwaters extended their overflow area, eroding the farmlands on the left bank upstream of the bridge for a while, and then crossed the road behind the abutment of the left bank of the bridge, furthering their overflow area to the farmlands on the left bank downstream of the bridge. Furthermore, although the overflowing floodwaters returned to the river at around 200 m downstream, the erosion of the farmlands in the overflow area probably continued until the water level dropped. In some cases, the topsoil of the paddy fields was eroded not only in the crop soil layer but also in the tillage and heart soil layers to a depth of about 3 m. Figure 6.18 is a photograph taken from the eroded farmland on the left bank upstream of the bridge where the river channel was blocked, looking toward the blocked bridge. It can be clearly seen that the ground behind the abutment on the left bank of the bridge has been washed away.

Figure 6.18 Damages on the left bank (Photo by Kenji Okajima).

6.5.2 *Ido River*

At the Ido River, the rising water washed away the headworks and eroded the dikes near the Ido water treatment plant in Kumano City. The overflowing floodwaters eroded the foundation of the water treatment plant, and sediments and debris carried by floodwaters filled rice paddies. As shown in Figure 6.19, the headworks were washed away at the survey location of the Ido water treatment plant. This headworks consisted of a fixed weir, which crossed the river at a slope of about 30° to the left bank, and streambed blocks to intake water from the left bank at the river's front line. The fixed weir on the right bank side of the headworks was washed away, and the revetment was heavily eroded. Although the fixed weir on the left bank remained, it had moved about 2 m downstream and slumped over to the right bank side. It is thought that the water level rose due to this headwork deformation, causing flood waters to overflow on both banks upstream of the headwork. As shown in Photo 6.6, a part of the bridge at the downstream end of the headwork was damaged by driftwood falling on the JR bridge.

6.5.3 *Kumano River and Onodani River*

The rising Kumano River caused flood sediments to fill rice paddies and inundate houses up to their second-floor roofs in the Asari district of Kiho

20m

Photographing
location

©2011 Google

Photographed on October 26, 2011

Figure 6.19 Damages near Ido River water treatment plant (Photo by Kenji Okajima).

Town. Furthermore, in the Onodani River Basin, where measures have been taken since 2001, such as constructing ring levees and raising residential lands (the Land-use Integrated Flood Disaster Prevention Project), in addition to the anticipated filling of paddy fields with flood sediment from the Onodani River, residential areas were also flooded due to floodwaters exceeding the projected high-water level flowing into the ring levees. In the Takaoka Area of Kiho Town, the rising water in the Onodani River broke the ring levee, flooding the area inside the levee, as shown in Figure 6.20 [8]. Moreover, the water level was high enough to move the roof tiles, as shown in Figure 6.21. Figure 6.20 and Figure 6.21 show the same area taken on September 5 and 4 respectively. Houses 1, 2, and 3 are the same houses in both photos. Comparing the two photos, it was found that the ring levee remained intact on September 4, but it overturned on September 5. This suggests that the ring levee failed because the water level outside the levee dropped, causing a massive volume of water to flow from the inside to the outside of the level. Thus, pressure was built on the levee, leading to its fall toward the outside. Because of this overturning failure of the ring levee, the foundations of the houses on the inner side of the dike were eroded, and the houses were severely deformed. In the future, efficient discharge of water within the levee should be examined at times of unexpected flooding when the causes of the ring levee failure are investigated.

Photo 6.6 Damages to a JR bridge (Photo by Kenji Okajima).

Photographed on September 5

Figure 6.20 Break point of a ring levee (Photo provided by Kiho Town).

The highest water level

(position of roof tile displacements)

9   Photographed on September 4

Figure 6.21 The position of the highest water level (Photo provided by Kiho Town).

6.6 Summary

The torrential rains of S.T.S. Talas in 2011 caused multiple sediment disasters and river disasters in the Chusei, Nansei, and Higashi-Kishu Regions of Mie Prefecture. Among them, large-scale slope failures exceeding 300 m in width and height occurred at Higashimata Valley in Odai Town and Kajiyamata Valley in Kihoku Town. The cause of such large-scale slope failures was likely an increase in the groundwater level because of the large amount of rainwater supplied through the water channels in the ground by long, continuous rainfall. In addition, many surface failures occurred in the Higashi-Kishu Region, from Kumano City to Kiho Town. Slope failures were concentrated around the south-facing slopes consisting of granite porphyry of the Kumano acidic igneous rocks. The relationship between rainfall and the locations of surface failures consisting of granite porphyry of the Kumano acidic igneous rocks suggests that a late onset of concentrated rainfall exceeding 80 mm/h in addition to a cumulative precipitation of 800 mm may have caused increased slope failures. Furthermore, damages to a World Heritage site, Kumano Kodō, due to river overflows in heavy rainfalls in the Higashi-Kishu Region and damages to the local population, such as damage due to the overtopping of a ring levee of the Onodani River, were observed.

References

[1] Disaster Countermeasure Headquarter, Mie Prefecture.: On Damages due to Severe Tropical Storm Talas of 2011 and Responses. Vol.47, 2011. <http://www.pref.mie.lg.jp/KOHO/talas/NO47.pdf.> (in Japanese, referenced 2015.1.5)

[2] Aizawa, T., Sakai, T. et al.: Sediment Disasters in Miyagawa Village, Mie Prefecture, Caused by Typhoon Meari of 2004, *Journal of the Japan Landslide Society*, 47(1), pp.26–33, 2010. (in Japanese)

[3] Geoinformation Service Center, The National Institute of Advanced Industrial Science and Technology: Geological Map Viewer GeomapNavi, 2014. <https://gbank.gsj.jp/geonavi/> (in Japanese, referenced 2014.6.18)

[4] Kawakami, Y. & Hoshi, H.: Ring Dikes and Sheet-like Intrusions in Volcanic-Plutonic Complex: The Kii Peninsula, Geology of the Kumano Acidic Igneous Rocks in Owase City – Kumano Region, *The Journal of the Geological Society of Japan*, 113(7), pp.296–309, 2007. (in Japanese)

[5] Suzuki, T.: *Introduction to Topographic Map for Construction Engineers*, Vol. 1, The Foundations of Map Reading, Kokon Shoin, p.122, 1997. (in Japanese)

[6] World Heritage "Sacred Sites and Pilgrimage Routes in the Kii Mountains" Three Prefecture Council (Mie, Nara, and Wakayama Prefectures): World Heritage "Sacred Sites and Pilgrimage Routes in the Kii Mountains" Preservation Management Plan, 2005. (in Japanese).

[7] Okajima, K., Ito, R., Kajisa, T., Yasuda, K., & Narioka, H.: Damages to Agricultural Lands and Facilities in Mie Prefecture Caused by the 2011 Severe Tropical Storm Talas, Journal of the Japanese Society of Irrigation, *Drainage and Rural Engineering*, 80(4), pp.40–46, 2012. (in Japanese)

[8] Photographs by Geospatial Information Authority of Japan: Aerial Photo ID Number CKK20113X, Photographed on 2011.9.6.

7 Slope Protection Measures in Japan and Restoration and Recovery Measures in the Kii Peninsula Disaster

Ryoichi Fukagawa

7.1 Introduction

As mentioned in Section 2.2, Japan is a country prone to sediment disasters. Followings are some of the causes of these disasters.

- Seventy percent of the country's land area is mountainous, with many slopes causing sediment disasters.
- The Japanese Archipelago exists in close proximity to a plate boundary and is constantly under horizontal pressure. As a result, it contains many geologically fragile layers such as faults, fracture zones, flow plates, and nodular structures, which can easily induce sediment disasters.
- The region is regularly hit by typhoons, which cause torrential rains, and there are periods of long-lasting rainfall associated with the onset of rainy season fronts and autumn rain fronts. The effects of typhoons and rainy season fronts or autumn rain fronts often overlap.

As described above, there are many causes that predispose or trigger sediment disasters. Slope protection measures must be taken with these factors in mind. Moreover, slopes can be artificial or natural, as described below in Section 7.2, and there are also various types of slope failures. This is characterized by the fact that slope disaster prevention measures must be diverse. Flexible measures must be taken according to the site conditions.

When applying slope disaster prevention measures, there are certain considerations that must be taken into account. Since there are countless slopes where sediment disasters could potentially occur, budget allocation must be prioritized. In other words, priority should be given to preventing further damage in the affected areas. The examples include check dam and bedding works in areas where debris flows have occurred, earth removal works on slopes where slope failure has occurred, and surface water and groundwater control works.

DOI: 10.1201/9781003375210-7

7.2 Slope Failure Protection Works in Japan

7.2.1 *Classification and Definition of Slopes and Sediment Disasters*

Normal slopes can be classified into artificial and natural slopes. Artificial slopes are defined as slopes of earth or rock artificially formed by fill or cut construction. On the other hand, when the term "slope" is used simply, it refers to the natural slope of the ground as it is. However, it is sometimes used in a broader sense that includes artificial slopes. These classifications are necessary in cases where slope disaster prevention measures must be modified depending on whether the slope is an artificial or natural slope. However, as noted in the previous section, natural slopes are the primary focus of this book, so descriptions of artificial slopes will be kept to a minimum in subsequent sections. Slope-related disasters can be divided into four categories based on the type of occurrence: Steep slope failure, landslide, debris flow, and rockfall [1]. For steep slope failures, they can be further classified into the following categories [1]:

   i Erosion and collapse
   ii Surface failure
  iii Large-scale slope failure
  iv Bedrock failure

First, "steep slope failure" is a general term for the rapid collapse of part or all of a slope due to the action of gravity. The four categories listed above are classified separately according to the type of slope failure, scale, and geology.

Erosion and collapse within steep slope failures include those where the surface is exfoliated or gullied by erosion from dry, wet, frozen, or surface water; those where overhanging portions of the slope collapse; and those where cracked or perturbed rock collapses.

Surface failure in steep slope failures includes those in which the topsoil slides down, those in which the surface layer of rock slides down due to weathering and so on, and those in which rock slides down along streambed structures or bedrock fractures. Surface failure is the most frequently occurring slope failure phenomenon in Japan, accounting for about 80% of all slope failures.

Large-scale slope failures within the category of steep slope failures include: Large-scale slides of slopes composed of soft, poorly consolidated strata or slopes with instability factors caused by rising groundwater; large-scale slides of rock bodies with geological structures such as flowbeds, faults, and fracture zones; and forward overturning of slopes of catch basins or rock with developed fractures. Figures 7.1 and 4.1 show a number of large-scale slope failure sites observed in southern Nara Prefecture during the Kii Peninsula Disaster. Akadani, Nagatono, and Ui (Shimizu), shown in Figure 7.1, were sites where disaster-prevention measures were essential because of the need for sediment damming measures or severe damage.

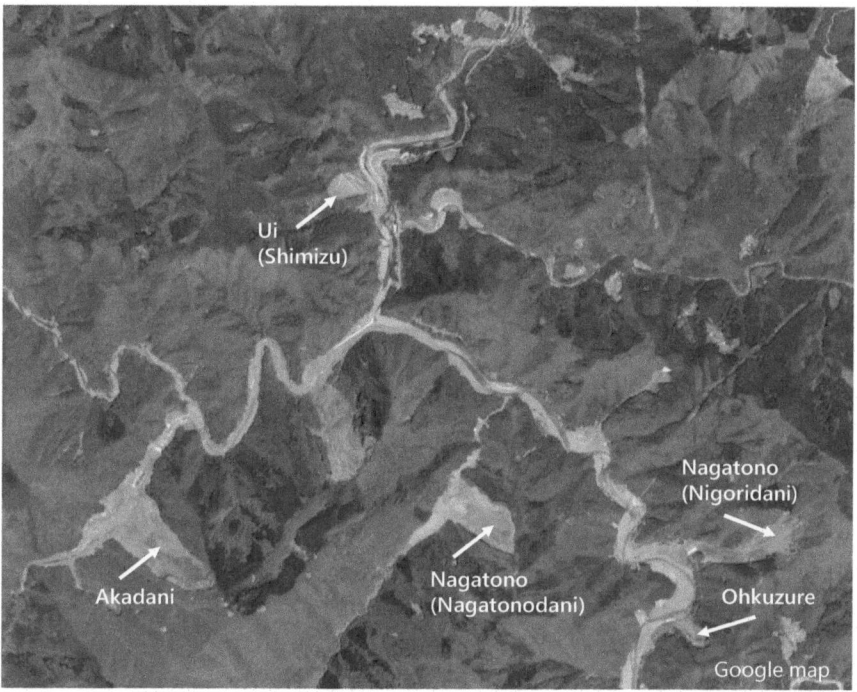

Figure 7.1 Large-scale slope failures at Gojyo City and Totsukawa Village (Created adding to Google Map).

For bedrock failures within steep slope failures, the term "bedrock failure" refers collectively to the phenomena of large-scale failure of rock bodies on bedrock slopes, including erosion and collapse above, surface failure, and large-scale slope failure.

A landslide is a phenomenon in which a certain surface deep underground serves as a boundary and the soil mass above it gradually moves downward. It tends to be concentrated in areas with specific geology and geological structures, and it causes large-scale movement of slopes with gentle gradients compared with steep slope failures, forming unique landforms. In Japan, it is known that landslides are common in areas along the Fossa Magna and the Median Tectonic Line, which are large-scale fault zones.

A debris flow is a phenomenon in which sediment, debris, and driftwood are fluidized by surface or groundwater and carried down a mountain stream, usually with great energy and destructive force. Debris flows have claimed the most victims in Japan's recent large-scale sediment disasters. Photo 7.1 shows many traces of debris flows observed in Nachikatsuura Town in southern Wakayama Prefecture during the Kii Peninsula Disaster. Some of the debris flows were several kilometers in length.

Photo 7.1 Aerial photo of Nachi River Basin just after disaster (Google Photo).

Rockfall is a phenomenon in which a rock mass or gravel is detached due to the expansion of fractures (nodular, schistose, or stratigraphic fractures that develop in the rock mass), and rocks, boulders, and gravels in cliff cone deposits, volcaniclastic debris, and poorly consolidated sand and gravel beds float to the surface and fall down from the slope.

As mentioned above, there is a wide range of slope-related disasters. Appropriate slope disaster prevention measures must be selected according to the anticipated disaster.

7.2.2 Outline of Slope Protection Works

Slope protection works are works to stabilize natural slopes in order to protect houses, roads, and so on from disasters caused by natural slope failure and so forth, or to improve those showing signs of disaster. The following items shall be taken into consideration when developing slope disaster prevention measures [1]:

  i Compatibility with intended use
 ii Safety of the structure
iii Durability
 iv Assurance of construction quality
  v Ease of maintenance and management
 vi Harmony with the environment
vii Economical efficiency

The basic policy based on the above considerations is as follows [1]:

1 If the scale of the anticipated disaster is large, the social impact will be great. Therefore, problem areas should be identified in the preliminary survey stage, and residential land development, routes, and road structures should be selected to avoid these areas as much as possible.
2 For the areas where measure works are to be carried out, the measure works should be carried out in accordance with the type and scale of the disaster, based on the ground and topographical conditions of the site.
3 Weaknesses should be identified and addressed even in the maintenance phase, and uncertainties such as heterogeneity of the ground and changes over time due to weathering and so on should be addressed.
4 Slope stabilization works for natural and artificial slopes should be designed to ensure safety and to take the surrounding environment and landscape into consideration as much as possible.

Slope protection works have the following peculiarities compared with ordinary works [1]:

1 Responding to the diversity of natural terrain and environmental considerations: The stability of natural and/or artificial slopes is complexly affected by rock quality, soil texture, depositional conditions, weather conditions, and other factors.
2 Importance of engineering judgment in the field: It is highly dependent on the experience and proper judgment of the engineer in the field. It is also based on the gradual improvement of slope performance.
3 Need to share information at each stage of investigation, design, construction, and maintenance: Uncertainties exist, such as heterogeneity of the ground and changes over time.

Slope disaster prevention works for natural slopes include slope failure measures, rockfall measures, rock collapse measures, landslide measures, and debris flow measures, depending on the type of slope-related disaster [1].

   Slope protection works can be broadly classified into two types: Control works and restraint works. Control works are methods of stopping or mitigating activities (increasing shear resistance) by altering natural conditions such as topography, soil texture, and groundwater. Restraint works are methods of restraint activity (i.e., obtaining slip resistance through other external forces) by means of a restraint structure.

Control works are used on a first-aid basis, and typical examples include the following works [1]:

1 Drainage work: This method is important because slope failures are mostly caused by groundwater.
2 Earth removal: All or part of the sliding soil mass, mainly near the head, is removed to stabilize the slope.
3 Slope protection: Typical methods include vegetation and structure-based protection.

On the other hand, restraint works include the list below. These works often require a long construction period and a large budget, and are implemented as a permanent measure [1].

1 Loading embankment method: Slopes are stabilized to resist sliding forces by placing fill at the end of landslides and slope failures.
2 Retaining wall method: Structures are built to prevent soil masses from collapsing or sliding against earth pressure, landslides, and other sliding forces.
3 Reinforced method: Embankments and soils are reinforced using geotextiles, reinforcing bars, and other reinforcement materials.
4 Landslide prevention pile method: This method to stop all or part of the landslide movement involves installing piles and using the resistance force of the piles.
5 Anchor method: A hole is drilled through the mobile layer (surface layer) of the slope to the immovable base, a steel material is inserted into the hole, and the tip of the material is fixed with grout. The anchor is then fixed to the anchorage plate on the ground surface, and the pull-out resistance and shear resistance of the steel material deters landslides and slope failures.

7.2.3 Measures for Slope Failure

Slope failure measure works can be roughly classified into preventive works and protective works. The preventive works are implemented to control the weathering and erosion of slopes and to prevent the occurrence of slope failures. The protective works, on the other hand, are executed to stop the movement of collapsed soil generated by the collapse or to change its direction to protect downstream settlements, roads, and passing vehicles.

The main preventive works include the following [1];

1 Wattle fence: This is a method of constructing fences as earth retention on mountainsides and riverbanks. Materials such as live plants, wood, steel, concrete, and so on, are used.

2 Precast frame construction: A type of crib work in which factory-made frame members are assembled on the slope surface, resulting in stable quality and excellent aesthetics. It has the function of preventing erosion caused by rainfall and other factors, and as a greening base, and is applied to gently sloping slopes.

3 Sprayed frame construction: A type of crib work in which a formwork is placed on the artificial surface and mortar is sprayed onto it. In addition to preventing bedrock avulsion and surface collapse, it also functions as a greening base.

4 Cast-in-place concrete framing work: This is a type of crib work in which formwork is placed on the slope surface and concrete is poured. This method is suitable for flat, low, artificial slope surfaces with large cross sections. It is mainly used for preventing bedrock avulsion and surface collapse, as well as for providing a foundation for revegetation. It is also used as a bearing structure for anchoring.

5 Retaining wall method: It is considered when stability cannot be ensured with a standard slope due to site limitations or topographical constraints in earthwork plans such as cut and fill and has the function of resisting earth pressure.

6 Continuous long-fiber reinforced soil method: This method minimizes the alteration of nature by using reinforced soil mixed with continuous long fibers by spraying, and the entire surface of the soil is revegetated by spraying vegetative substrate and so on. This method is used in combination with the anchoring method and soil reinforcement method to achieve the dual objectives of stabilizing the ground and improving the landscaping.

7 Pile works: This method should have a relatively large restraint force and should be constructed on slopes where the base is strong enough to accommodate the moving soil mass. A typical application example is pile groups used for landslide movement.

8 Ground anchors method: This method is used to provide restraint force on cut surfaces of relatively compacted soil or on natural slopes where there is a risk of collapse. A ground anchor consists of an anchor body, a tension section, and an anchor head.

9 Soil nailing method: This method improves the stability of the ground pile by supporting earth pressure, stabilizing slopes, and increasing bearing capacity by placing reinforcement that does not pre-stress the ground pile and by passively exerting a resisting force on the reinforcement as the ground deforms, thereby restraining ground deformation.

The above measures are commonly used in ordinary construction projects. Figure 7.2 [2] shows an overview of the continuous long-fiber reinforced soil method (the Geo-Fiber Method). The main feature of the Geo-Fiber Method is that the slope is protected by continuous fiber-reinforced soil, which is a mixture of sandy soil and fibers. Mixing continuous fibers into sandy soil imparts pseudo-consistency to the sand, resulting in a reinforced soil with

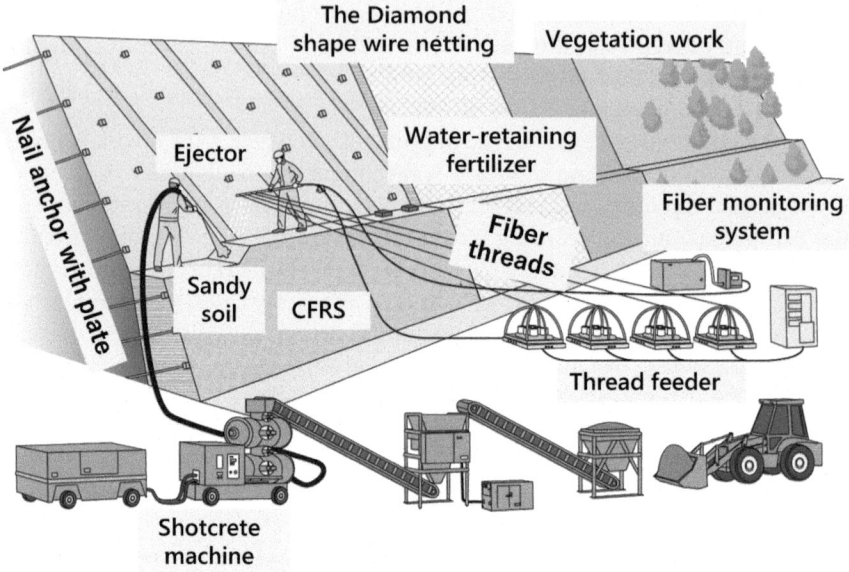

Figure 7.2 Continuous long-fiber reinforced soil method: the Geo-Fiber Method [2].

high shear strength and tenacity. This tough, continuous, fiber-reinforced soil protects the slope surface. The surface of the continuous fiber-reinforced soil can be vegetated to create lush landscapes and natural environment-friendly slopes, was not possible with conventional block retaining walls or mortar/ concrete slopes.

Figure 7.3 shows the outline of the Non-Frame Method [3]. Until now, slope failure measures have consisted of covering the slopes with concrete structures. In contrast, the Non-Frame Method is a revolutionary slope failure prevention technology that does not cut down trees growing on the slope and preserves the original landscape and environment after construction. Steel rods (rock bolts) are driven into stable ground at a depth of approximately 2 to 3 m below the surface, and steel plates (support plates) are attached to the ground surface to hold down the collapsible soil. These are placed on the entire slope at about 2-meter intervals, and the entire slope is further connected with wire ropes to protect the slope. Components and construction methods have been developed to allow construction without cutting down trees on slopes, so that the original landscape and natural environment can be maintained after construction.

These two construction methods are known as landscape-friendly construction methods and have many construction examples. They are particularly important in areas where scenery is important, such as sightseeing spots.

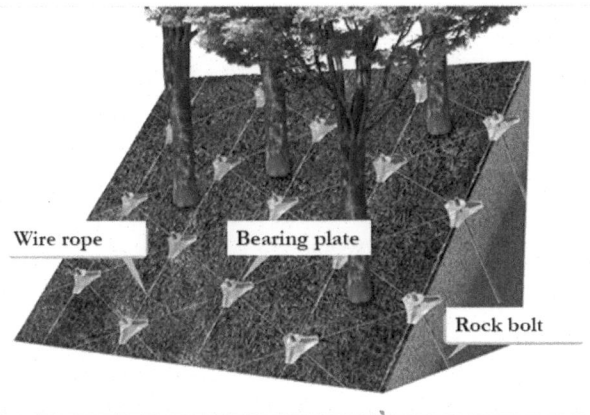

Figure 7.3 The Non-Frame Method [3].

In addition, the main protective works include the following [1]:

1 Waiting retaining wall method: This method is employed when there is sufficient space available at the back of the retaining wall.
2 Soil-covering works: This method is used when the expected amount of collapsed soil is large, or when the expected location of collapse is high and the impact force of collapsed soil is large, or when the slope on the mountain side of the road is too steep to accommodate the pocket capacity of collapsed soil.

7.2.4 Measures for Rockfall

Rockfall prevention works consist of works to prevent rockfall at the source and rockfall protection works to mitigate damage caused by falling rocks. The principles of rockfall measures include the following items [1];

 i To prevent weathering erosion that causes rockfall
 ii To stop the generation of falling rocks
 iii To absorb energy of falling rocks
 iv To alter the direction of falling rocks and direct them harmlessly
 v To resist impact and stop falling rock movement
 vi To combine the effects of preventing small-scale collapsed earth from falling and avalanche prevention

The functions, durability, workability, economic efficiency, and maintenance and management issues of various measure works must be carefully examined to select the most appropriate type and combination of works for the local road and/or village conditions.

Rockfall prevention work is a source control measure implemented for floating and boulders on slopes where rockfall is expected to occur, with the expectation that it will have the following effects [1]. Typical construction methods corresponding to each effect are also shown.

1 To prevent falling rocks due to erosion and weathering: Adhesive construction method
2 To secure individual rockfalls: Foot protection works
3 To remove fallen rocks: Stone removal work
4 To deter rockfalls in their entirety: Rockfall protection net (+Rock-bolt anker)
5 To prevent rockfall due to slope failure: Spraying method (+Rock-bolt anker)

On the other hand, rockfall protection works are waiting measures to protect falling rocks from slopes by installing facilities in the middle of slopes, at the roadside, or on the road. The types of rockfall protection works are classified according to the location where they are installed as follows [1]. Typical construction methods are also shown below.

1 In the middle zone (on the slope) between the source and the road: Rockfall protection fences, rockfall protection nets, rockfall protection walls.
2 On the roadside (at the bottom of the slope): Rockfall protection fences, rockfall protection nets, rockfall protection walls, rock sheds, and rockfall protection embankments

7.2.5 *Measures against Bedrock Collapse*
There are three approaches to rock collapse prevention [1].

  i Measures by avoidance
 ii Measures by physical measure works
iii Temporary measures by monitoring

Since it is difficult to estimate the mechanism of rock failure, it is desirable to refer to past cases or conduct as detailed a survey as possible when selecting and implementing specific measures. Considerations in the selection of construction methods include the following items [1]:

1 The measure method is assumed to be the same as that for rockfall measures.
2 However, due to the scale of the subject and the large destructive force, careful investigation and analysis will be conducted to determine the avoidance of rock slopes.

3 Due to the large number of unknowns, it is desirable to avoid as much of the collapse reach of the rock slope as possible.

7.2.6 Measures for Landslide

Landslide protection works are roughly classified into control works and restraint works. The control works are methods to stop or mitigate landslide by changing natural conditions such as topography and groundwater conditions. The restraint works stop a landslide in whole or in part by the restraint force of the structure.

Control works can be further classified as follows [1]:

1 Surface-water drainage works: This method eliminates infiltration of precipitation into the landslide area and seepage water from ponds, swamps, and so on. It is effective in many cases as a landslide measure.
2 Underground-water drainage works: If groundwater is affecting the landslide, groundwater drainage works should be performed. Depending on the depth of the groundwater zone, there are shallow groundwater drainage works and deep groundwater drainage works.
3 Groundwater insulation method: This method intercepts groundwater inflow along a well-defined flow path from outside the landslide block.
4 Earth removal works: This method increases stability by eliminating all or part of the landslide block. Usually, a portion of the upper half is eliminated.
5 Loading embankment method: The landslide is stabilized by filling the landslide terminus with earth masses.
6 Control by river structures (weirs, bedding works, water control works, revetment works).

Riverbed degradation and streambank (rifflebank) erosion caused by running water may compromise the stability of the landslide soil mass. In such cases, we aim to protect the stream bank and stabilize the landslide terminus.

Typical restraint methods include the following prevention pile method and shaft construction.

1 Prevention pile method: Install large-diameter boreholes at a predetermined depth below the slip surface. Boring shall be done vertically by cutting the slip surface, and steel pipes, and so on, shall be inserted. Generally, piles should be installed in a position where the soil reaction force due to the soil mass behind the pile (valley side) is sufficient to maximize the strength of the pile.
2 Shaft construction: When general pile construction is not feasible, a method of constructing a reinforced concrete shaft structure can be employed by digging a well with a diameter of 2.5 m to 6.5 m instead of large-diameter boreholes.

7.2.7 Measures for Debris Flow

When targeting roads for debris flow protection, small shifts in routes and road structures, such as culverts, should be considered first. If the road structure alone is unable to cope with the problem, consideration should be given to using check dam or other debris flow measure facilities. However, there may be cases where large natural slopes cannot be handled by measure works alone. In such cases, efforts should be made to ensure the safety of road traffic by utilizing means such as traffic restrictions [1].

At the site of a disaster, the village is the object of conservation, and it is often difficult to move the village itself. Therefore, in order to prevent debris flows and driftwood, the following measures should be kept in mind: (1) It is important to prevent debris flows and driftwood from occurring; (2) when debris flows and driftwood occur, they are not allowed to reach the village but are dealt with along the way; (3) when debris flows and driftwood occur, the flow route is changed so that they do not affect the village.

Specific measures against debris flows and driftwood are listed below [4,5].

1 Debris flows and driftwood control works: These works consist of hillside works to control the generation of debris flows and driftwood in water source areas, check dam, bedding works, strip works, revetment works, stream-bank protection works, and so on, to prevent sediment movement in streambeds.
2 Erosion control facilities that capture debris flows and driftwood in a standby manner. Typical examples are check dam and driftwood trapping facilities.
3 Debris flow diversion works: This is a method of safely diverting runoff to downstream areas with little land use, without depositing it in the middle of the section to be preserved.
4 Debris flow deposition method: This method involves widening or gently sloping the flow channel, which includes a debris flow dispersal deposition area and a debris flow deposition channel.
5 Debris flow buffer zone: In this method, a zone of trees is established for the purpose of reducing the energy of debris flows.
6 Directional control of earth and rock flow: This method controls the direction of earth and rock flow by means of a dike or other object.

Among the above, check dam, stream protection works, and bedding works are frequently used as basic construction methods in the rehabilitation and reconstruction projects of the Kii Peninsula Disaster, so they are briefly introduced here. First, check dam can be further classified into permeable, partially permeable, and impermeable types. Bedding works are facilities installed across a river to prevent scouring of the riverbed and stabilize the river gradient. Those with a drop-off are called "stepped works" and those without a drop-off are called "bed sill." The height of a drop-off structure is usually less than 5 m. They are often used in conjunction with revetments. Stream-bank protection works are planned and constructed to safely carry away sediment and floodwaters, and to avoid altering the existing stream as much as possible, thereby ensuring flood control safety and preserving the stream's ecosystem. Stream-bank protection works are basically composed of facilities to control turbulence of floodwaters and excessive fluctuation of streambed height (bed stabilization works, strip works, etc.), facilities to prevent stream-bank erosion (revetment works, water control works, etc.), and facilities to deposit fine sediments (riparian forests, etc.).

Photos 7.2 and 7.3 show impermeable and partially permeable check dam installed in the Narukotani River at a large-scale debris flow site in Nachikatsuura Town, Wakayama Prefecture, respectively. Many similar check dam were installed in the affected areas of Nachikatsuura Town.

Photo 7.4 shows the stream protection works installed on the Higuchi River, also in Nachikatsuura Town. It can be seen that it consists of a drop-off structure, a bedding structure, and a revetment structure, which is a type of bedding structure.

Photo 7.5 shows a typical sand pocket area, which was established along the Kanayama-dani River in the town of Nachikatsuura.

Photo 7.2 Typical example of impermeable check dam used in Kanayamadani River, Nachikatuura Town (Photo by Ryoichi Fukagawa).

Photo 7.3 Typical example of partially permeable check dam used in Kanayamadani River, Nachikatuura Town (Photo by Ryoichi Fukagawa).

Photo 7.4 Typical example of stream protection work, combination of ground sill and revetment, used in Higuchi River, Nachikatsuura Town (Photo by Ryoichi Fukagawa).

Photo 7.5 Excavation and hauling operation using unmanned construction carried out at Akadani Area [7].

7.3 Slope Rehabilitation and Reconstruction Project in the Kii Peninsula Disaster

7.3.1 Slope Rehabilitation and Reconstruction Project in Nara Prefecture

7.3.1.1 Measures after a Major Collapse in the Akadani Area of Totsukawa Village [6–8]

At the time of the disaster, the Akadani area suffered a collapse 460 m wide, 600 m high, and 850 m long. The collapsed sediment, amounting to approximately 11.38 million m^3, blocked the river channel and formed a natural dam, which has now been reclaimed (see Figure 7.1). Although the risk of overflow and collapse has decreased due to the reclamation of the flooded pond, secondary movement of sediments in the blocked channel and unstable sediments in the collapsed area may cause extensive damage downstream in the Shimizu, Nagatotono, Umiyahara, and Uenoji Areas. The work was undertaken to prevent the discharge downstream of sediments deposited in the channel blockage and unstable sediments within the collapsed area. Figure 7.4 shows the current status of the Akadani Area. As of September 2022, weir No. 3 has already been completed upstream of weir No. 2. In this figure, the stream protection works were implemented to improve safety by preventing sediment runoff from channel erosion and direct erosion damage to downstream preservation targets such as camp sites. The ground sill works were implemented to scrape the planned riverbed to the original bed and to improve safety by preventing sediment runoff due to channel erosion.

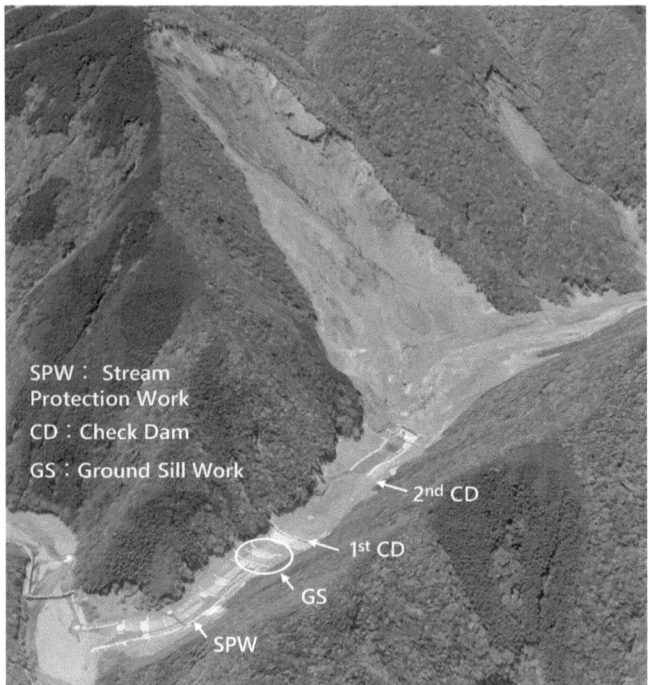

Figure 7.4 Disaster mitigation measures at Akadani Area (Created adding to Google Earth).

In the Akadani Area, even after the post-disaster collapse measures were completed, collapses from the remaining fallen portions frequently occurred, especially during the flood season. When recollapse occurred during typhoons, significant sediment deposition occurred, and even when recollapse did not occur, channel erosion took place, which had a significant impact on the construction of measure works. Such damage occurred six times in major cases alone. Therefore, unmanned construction was attempted in many cases where the safety of the construction workers was problematic. Photo 7.5 shows excavation and hauling operations using unmanned construction during the runoff period. There were times when up to 15 construction machines were in operation. By September 2022, the series of disaster-prevention measures moved to the final phase, the No. 3 check dam construction. This construction work was to be done in the area closest to the collapsed slope, so ensuring safety was a top priority. Therefore, human access was prohibited during the flooding season (June 15 to October 31). In order to complete the construction work early, in addition to unmanned construction by remote control during the outflow period, a new automated construction method was proposed and implemented for the first time in Japan for erosion control construction. In general, unmanned construction is operated remotely while the operator checks the monitor, which is said to reduce construction efficiency by about 70% compared with normal manned construction. Kajima Corporation implemented automated construction for the placement of concrete blocks and soil cement spreading and compaction work in the construction of a check dam, identifying tasks that were time-consuming and difficult to perform with remote construction while watching a monitor.

7.3.1.2 Measures after a Major Collapse in the Nagatono Area of Totsukawa Village [6,7]

At the time of the disaster, a 340 m wide, 400 m high, and 650 m long large-scale slope failure occurred in the Nagatono Area (see Figure 7.1). The collapsed sediment, amounting to approximately 5.95 million m^3, blocked the river channel, creating a waterlogged pond. Although many large-scale slope failures have occurred in Nara Prefecture, this is the only affected site where freshwater areas still remain. When heavy rainfall occurs, the over-flow from the flooded pond causes debris flows due to rapid erosion of the channel-clogging sediment, which may cause extensive damage in the downstream areas of Nagatono, Umiyahara, and Uenoji. Work was un-dertaken to prevent rapid erosion of the channel-clogging sediment and to prevent the discharge of sediment deposited downstream of the channel-clogging area. Figure 7.5 shows the situation. In this figure, a drainage tunnel was installed to lower the water level of the flooded pond (putting in a culvert drainage pipe using the pipe jacking method), as well as to safely channel the floodwaters and reduce the risk of erosion of the embankment. Driftwood block work also plays a role in preventing blockage of channel works by driftwood.

7.3.1.3 Measures after a Large-Scale Slope Failure in the Ui (Shimizu) Area of Gojo City [6,7]

In Nara Prefecture, there have been many so-called large-scale slope failures. One of them was a large-scale slope failure in the Ui (Shimizu) Area of Gojo City. In this slope failure, clods of earth from the massive collapse crossed the river in front of the slope and pushed into the village on the other side of the river. As a result, 11 people were killed. This was the largest damaged site in Nara Prefecture. Although the amount of collapsed sediment was not so large (see Figure 7.1), immediate disaster-prevention measures were taken because houses remained in the vicinity of the collapsed slope. At the time of the disaster, the Ui Area had collapsed 220 m wide, 250 m high, and 350 m long, generating approximately 1.6 million m^3 of collapsed sediment. Emergency work was completed to stabilize the collapsed slopes and to improve the revetment for the safe flow of floodwaters. Figure 7.6 shows the current sit-uation of the area. In this figure, the revetment was implemented to prevent erosion of the riverbanks and stabilize the slope.

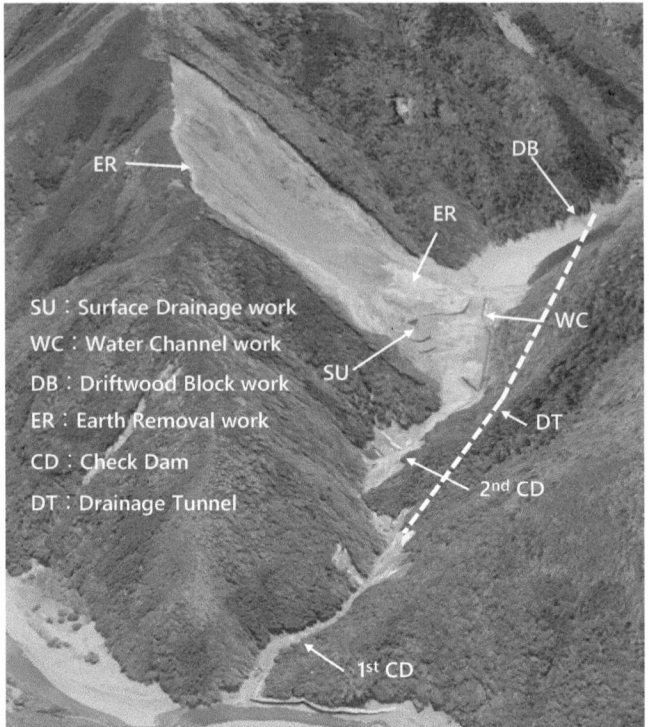

Figure 7.5 Disaster mitigation measures at Nagatono Area (Created adding to Google Earth).

7.3.2 Restoration and Reconstruction Projects in Wakayama Prefecture

7.3.2.1 Debris Flow Measures in the Nachi River Basin [6]

In the Nachi River Basin, many slopes collapsed in the tributaries flowing into the Nachi River at the time of the disaster, and the debris flows caused a great deal of human suffering and property damage (see Photo 7.1). In the entire basin, 27 people were killed, 1 person is missing, 103 houses were completely destroyed, 905 houses were partially destroyed, 440 houses were inundated above floor level, and 962 houses were inundated below floor level. As of now, a total of 15 check dams have been completed in eight tributary rivers as part of emergency measures. Figure 7.7 shows the construction work that has been done to date. There are elementary schools and private houses downstream of each branch of the river, and many residents still live in the area. This is a district where priority disaster-prevention measures have been implemented.

The specific construction method was based on a combination of check dams and stream protection works. In some tributary streams, weirs

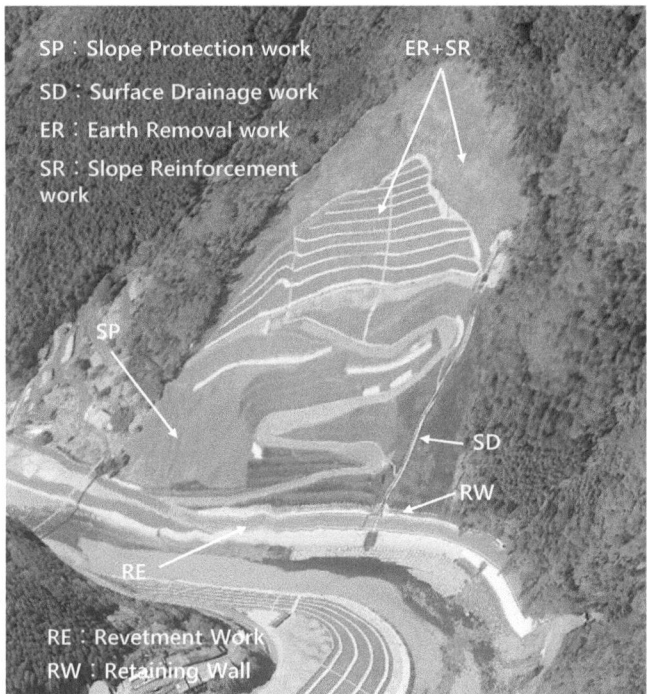

Figure 7.6 Disaster mitigation measures at Ui (Shimizu) Area (Created adding to Google Earth).

were constructed in double layers. For the main Nachi River and the Kanayamadani River, one of the tributaries, a sandy pocket was established respectively, due to the large scale of the debris flow.

7.3.3 Restoration and Reconstruction Projects in Mie Prefecture

7.3.3.1 Measures after a Large-Scale Slope Failure in the Higashimata Valley Area [9]

The amount of collapsed sediment in the Higashimata Valley is estimated to have exceeded 1,950,000 m³, especially in the section from the collapse site to the confluence of the tributaries, where unstable sediment of 90–150 m in width and 25–35 m in depth was deposited. There was concern that this unstable sediment would be washed downstream as sediment during heavy rainfall events. Therefore, Mie Prefecture plans to implement measures as shown in Figure 7.8 by 2025. A small check dam based on INSEM (In-situ Stabilized Excavated Material Method) was installed at the foot of the collapsed area to control riverbed erosion. Downstream of the channel, a steel channel construction (L = 273.3m) was installed to safely remove flood waters and fix the water route. In addition,

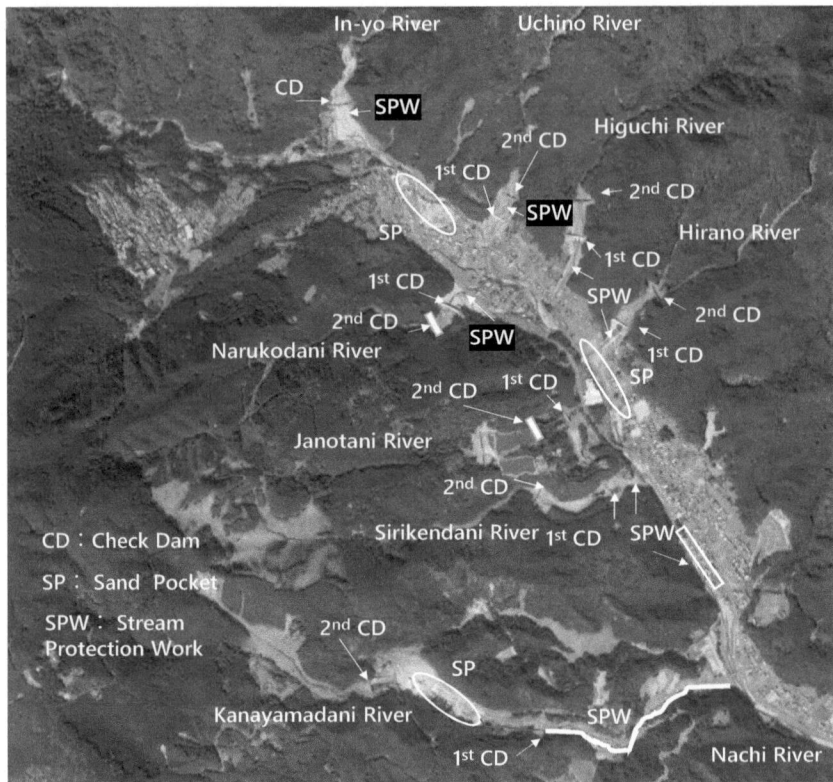

Figure 7.7 Disaster mitigation measures at Nachi River Basin (Created adding to Google Earth).

bedding works were constructed to anchor unstable sediments to the site and control sediment discharge. Three concrete small check dams are planned to be installed downstream of the confluence to prevent the discharge of unstable sediment. At the downstream end, a steel slit dam was installed as a measure against driftwood by extending the existing small check dam, thereby reducing the risk of disasters to downstream living areas.

7.3.3.2 Measures after a Large-Scale Slope Failure in the Kajiyamata Valley Area [9]

A large amount of sediments and driftwood that moved from the collapsed Kajiyamata Valley were still unstable and deposited inside the failed area and on the riverbed. These unstable sediments were of concern because of the possibility of runoff in future rainfalls; thus, Mie Prefecture installed three check dam to prevent the runoff of the unstable sediments upstream, as shown in Figure 7.9.

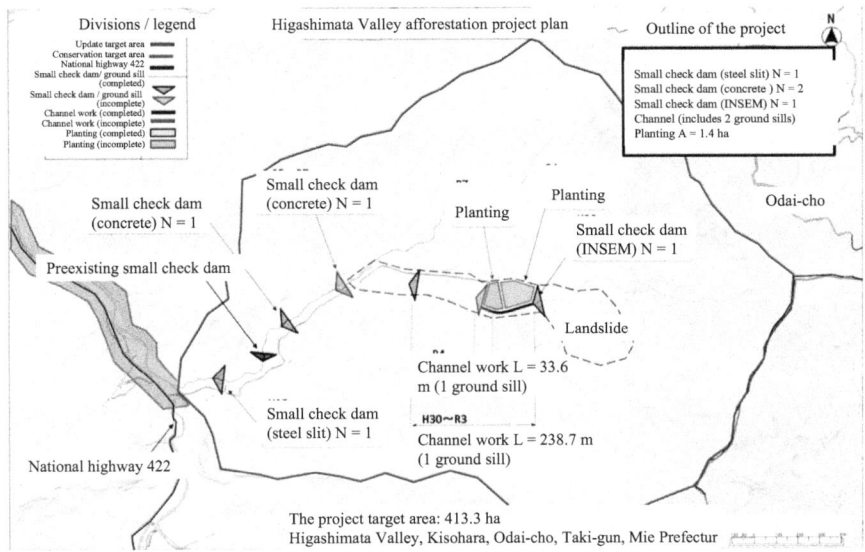

Figure 7.8 Disaster mitigation measures plan at Higashimata Valley Area (Original map provided by Mie Prefecture).

7.3.4 *Cultural Heritage Restoration and Reconstruction Projects [10]*

"Sacred Sites and Pilgrimage Routes in the Kii Mountain Range" was registered as a World Heritage site in 2004 in recognition of the three sacred sites and their cultural landscapes: Kumano Sanzan (Three Kumano main thrines) for Kumano worship, Yoshino, Omine for Shugendo (mountain asceticism), and Koyasan for Shingon esoteric Buddhism. This vast World Heritage site, which stretches across the Kii Peninsula, has frequently suffered sediment disasters due to its steep terrain and annual precipitation exceeding 3,000 mm In the 2011 Kii Peninsula Disaster, sediment disasters occurred in many places. In particular, a large-scale slope failure, a surface failure, and a debris flow occurred at Iseji-Yokogaki Pass, one of the pilgrimage routes, respectively, resulting in the loss of about 250 m of the pass in total at three locations. Disaster-prevention measures were taken in response to these disasters, and we introduce here the restoration and reconstruction measures taken in response to the large-scale slope failures.

The large-scale slope failures consisted of a collapse that occurred in two directions near the saddle between Mt. Nishinomine and the mountain to the south of it (elevation 316 m), and a collapse that occurred to the north of the saddle. Although a portion of the pass remained, the ground was washed away in an area approximately 95 m wide, 200 m from the top of the cliff to its end, and a 60 m height difference, making it impossible to restore the pass at the same location and elevation. Since the forest road could not be restored

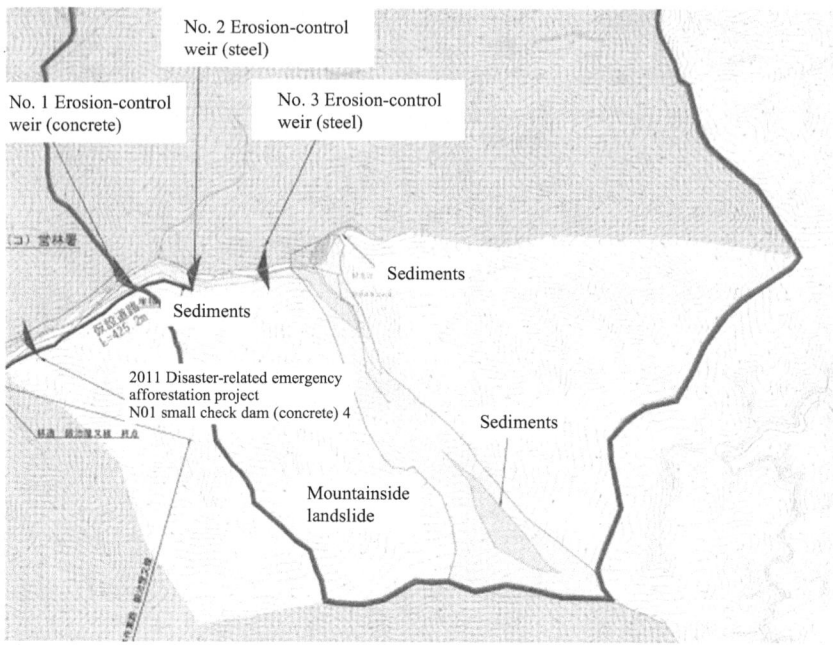

No. 2 Erosion-control
weir (steel)

No. 1 Erosion-control
weir (concrete)

No. 3 Erosion-control
weir (steel)

Sediments

Sediments

2011 Disaster-related emergency
afforestation project
N01 small check dam (concrete) 4

Sediments

Mountainside
landslide

Figure 7.9 Disaster mitigation measures plan at Kajiyamata Valley Area (Original map provided by Mie Prefecture).

to its original location, the slope on the south side of the mountain was cut and a site was secured. The cut slope was sprayed with colored concrete mixed with the color of the surrounding soil in consideration of the landscape. The cliffs were sprayed with a thick layer of base material near the top of the slope and with colored concrete in all other parts of the slope. Two stone-textured small check dams were constructed near the bottom of the slope and downstream of the slope. The remaining part of the slope was kept as it was, and the slope was formed by the collapsed sediments upstream to the forest road and downstream to the top of the small check dam. The slope was planted with trees, a futon cage was placed at the bottom of the slope, and a surface drainage channel made of fabric formwork was installed from the forest road upstream to the top of the valley drainage channel. Figure 7.10 shows a distant view of the collapsed area.

Although the landscape, including the buffer zone, has deteriorated because of the collapse and remedial work, it is expected that the landscape will improve with the growth of tree vegetation after several decades. When the slope collapses, the pre-disaster landscape is lost, but although the ground and trees are washed away, it creates space in the mountains and opens up a view from the cliff, which may be used for tourism and disaster education. How to make use of heritage that cannot be restored to its original state is

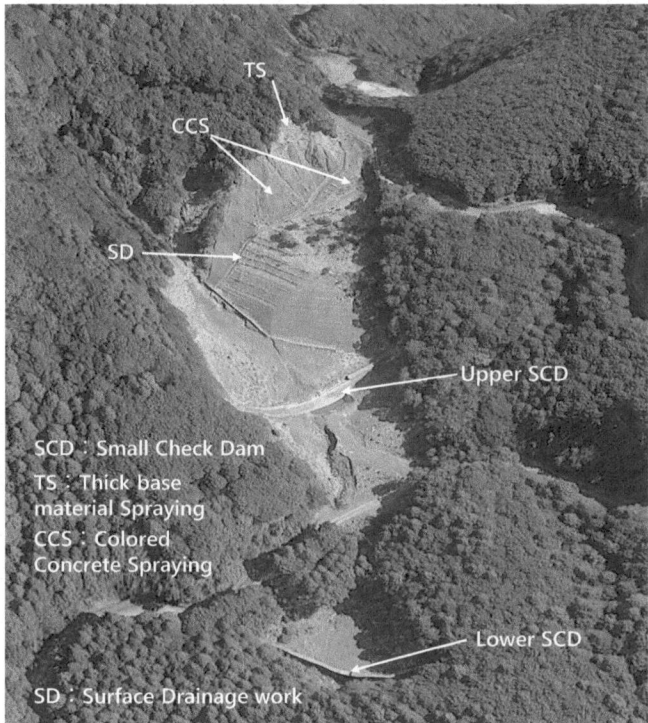

Figure 7.10 Disaster mitigation measures at Yokogaki-Tohge Pass of Kumano-Kodo, Mie Prefecture (Created adding to Google Earth).

also an issue encompassed in cultural landscapes. It is necessary to consider how we who live in the modern age can relate to and inherit the great nature of the Kii Mountain Range in our daily lives, and the paths of prayer that have continued since ancient times, through disasters.

7.4 Summary

This chapter first provides an overview of slope failure protection works in Japan, followed by a region-by-region introduction of typical rehabilitation and reconstruction projects in the Kii Peninsula Disaster.

Slopes are broadly classified into artificial slopes and natural slopes, but this chapter mainly focuses on natural slopes. Slope hazards are also classified into steep slope failure, landslide, debris flow, and rockfall based on the type of occurrence. Among these, steep slope failure, which occurs most frequently, is further classified into erosion/collapse, surface failure, large-scale failure, and bedrock failure. Because of the wide variety of slope failure phenomena, there must also be a wide variety of slope disaster prevention measures. In this chapter, measures for slope failure, rockfall, bedrock collapse, landslide, and debris flow are discussed, and their outlines and representative measures are introduced.

Representative rehabilitation and reconstruction projects in Nara, Wakayama, and Mie prefectures were introduced, with particular emphasis on projects in areas affected by large-scale slope failures and large-scale debris flows, which had a significant impact on each region. While disaster-prevention measures have generally been completed in many areas, there are still some places where measures are ongoing. In some cases, such as in de-populated areas, the implementation of the project itself has been postponed. The declining birthrate and aging population are particularly noticeable in depopulated areas, and it is expected that more and more places will have to abandon restoration and reconstruction in the future.

References

[1] Japan Road Association: Guidelines for Road Earthworks, Earth Cutting and Slope Stabilization (2009 Edition), 2018. 11. (in Japanese)

[2] Ujihara, M. and Fukagawa, R.: The Geo-Fiber Method-Protecting slopes with environment-conscious fiber reinforced soil-, The Fifth World Landslide Forum (Kyoto), Theme 6 Specific Topics in Landslide Science and Applications, Session 6.E2 Introduction of landslide mitigation measures of Japan (E-Proceedings), 2020. 11.

[3] Non-Frame Construction Method Association: Pamphlet for non-frame construction method <http://www.non-frame.com/about/structure.html> (referenced on 11th Feb.)

[4] National Institute for Land and Infrastructure Management, Research Center for Disaster Risk Management, Erosion and sediment control division: Manual of Technical Standard for Establishing Sabo Master Plan for Debris Flow and Driftwood, *Technical Note of National Institute for Land and Infrastructure Management*, No.364, 2007. 3. (in Japanese)

[5] Japan Society of Erosion Control Engineering: Dictionary of Technical Terms for Sabo Engineering, 2004. 12. (in Japanese)

[6] Sabo office of Kii Mountain Range: Sabo projects under direct control of Ministry of Land, Infrastructure, Transport and Tourism <https://www.kkr.mlit.go.jp/kiisankei/map/> (in Japanese, referenced 2023/1/8)

[7] Based on construction record provided by the Sabo office of Kii Mountain Range (in Japanese)

[8] Kajima Corporation: Unmanned construction in Akadani <https://www.kajima.co.jp/news/digest/ jul_2022/feature/03/index.html> (in Japanese, referenced 2023/1/8)

[9] Based on information provided by the Mie Prefecture (in Japanese)

[10] Ishida, Y.: Efforts in Kumano Pilgrimage Routes, *Case Studies on Slope Assessment and Measures for Disaster Mitigation of Cultural Heritage*, No. 2, Institute of disaster mitigation for urban cultural heritage, Ritsumeikan University, pp.12–22, 2023. 3. (in Japanese)

8 Disaster-Prevention and Mitigation Measures Following the Kii Peninsula Disaster

Kazuaki Hioki

8.1 Introduction

In 2011, the record-breaking torrential rain of the Severe Topical Storm Talas caused large-scale sediment and flood disasters across the Kii Peninsula, especially in the Nara, Wakayama, and Mie prefectures. Since then, large sediment disasters and river disasters have continued to occur across Japan, making us realize that the severity of natural disasters has increased in recent years. In general, updating disaster-prevention infrastructure (hardware measures) is of great importance for effective disaster prevention and mitigation; however, as many disasters are caused by external forces that go beyond the design assumptions, effective software measures (such as self-help and mutual-help disaster-prevention skills by community residents and technologies to predict disasters and convey necessary information to those who need it) have become essential today.

This chapter introduces disaster-prevention and mitigation measures focusing on software measures, beginning with the case of the Kii Peninsula Disaster. Specifically, we introduce a system to monitor the risks presented by deep-seated landslides in Totsukawa Village, Nara Prefecture; flood and inundation measures by local governments such as Shingu City in Wakayama Prefecture; and the timeline implemented ahead of the rest of the country by Kiho Town in Mie Prefecture. In addition, we introduce a slope surface monitoring system for highways under the jurisdiction of the West Nippon Expressway Company, as well as relevant measures undertaken by the West Japan Railway Company according to railway operation regulations. These measures are limited in terms of content and areas, but we hope that they contribute to future disaster-prevention and mitigation measures.

8.2 Disaster-Prevention and Mitigation Measures Based on the Kii Peninsula Disaster

8.2.1 *Monitoring Deep-Seated Landslide Risks in Totsukawa Village*

The record-breaking torrential rain of the S.T.S. Talas (the Kii Peninsula Torrential Rain) in 2011 caused many deep-seated landslides in the southern

DOI: 10.1201/9781003375210-8

part of Nara Prefecture. In this area, deep-seated landslides mainly occurred on the slopes of sedimentary rocks with accretionary complexes. Although very few cracks exist in the upper part of the rocks, some rocks on these slopes had well-developed cracks. This is considered to have likely caused easy storage of rainwater, which permeated from the surface layer in these rocks with developed cracks. In this area, another large river flood, the Meiji Kii Peninsula Torrential Rain Disasters in 1889, had caused sediment disasters. The recent climate trend has suggested an urgent need to develop and operate a specialized monitoring system for deep-seated landslide risks during torrential rains in this area.

One of the methods used to predict landslide risks during torrential rains is the hydrological method, which focuses on groundwater content. This method assumes that "rainfall-induced landslides would occur when the water content in the ground reaches a certain value," and the associated storage limit is obtained using a tank model. A typical example is the soil water index of the Japan Meteorological Agency [1]. The soil water index predicts the landslide risk from the groundwater content obtained from a hypothetical three-stage tank model with specific parameters (the sum of the storage volume for a three-stage tank). However, it fails to predict the landslide risk of each location considering the impact of the landform, geology, and vegetation. Nevertheless, as the soil water index can reflect the impact of the previous rainfall, it plays an important role in setting the reference rainfall for landslide warnings and evacuation. It is also used as evaluation criteria for municipality mayors to issue evacuation advisories and instructions to their residents at an appropriate time. In general, the soil water index mainly applies to relatively shallow ground. Therefore, it cannot be used to predict the risk of deep-seated landslides, such as the one that occurred in the southern part of Nara Prefecture in 2011 [2].

Following the Kii Peninsula Disaster, Hioki et al. (2018) [3] analyzed a deep-seated landslide that occurred in the southern part of Nara Prefecture in 2011 using a unique tank model (the Aoki-Hioki model; Figure 8.1), in which the permeability changes with the storage volume from the upper tank to the lower tank. They proposed an index for the risk of deep-seated landslides caused by torrential rains in the same area (Figure 8.2) and monitored the deep-seated landslide risk for Totsukawa Village in Nara Prefecture. The proposed index focused on the second tank of the model and used its storage volume as the index for the deep-seated landslides risk index, hereafter abbreviated as LRI. In addition, the index predicted the risk from the hypothetically obtained water content of the rocks, although it could not predict the risk for each location that took into consideration the impact of landforms and geology. The 2011 deep-seated landslide occurred when the LRI and its one-week cumulative value were ≥280 and ≥23,000, respectively. The zones that satisfied these conditions were classified as deep-seated landslide occurrence zones (red zone). The zones with a LRI of ≥280 and a one-week cumulative value of <23,000 were classified as warning zones (yellow zone).

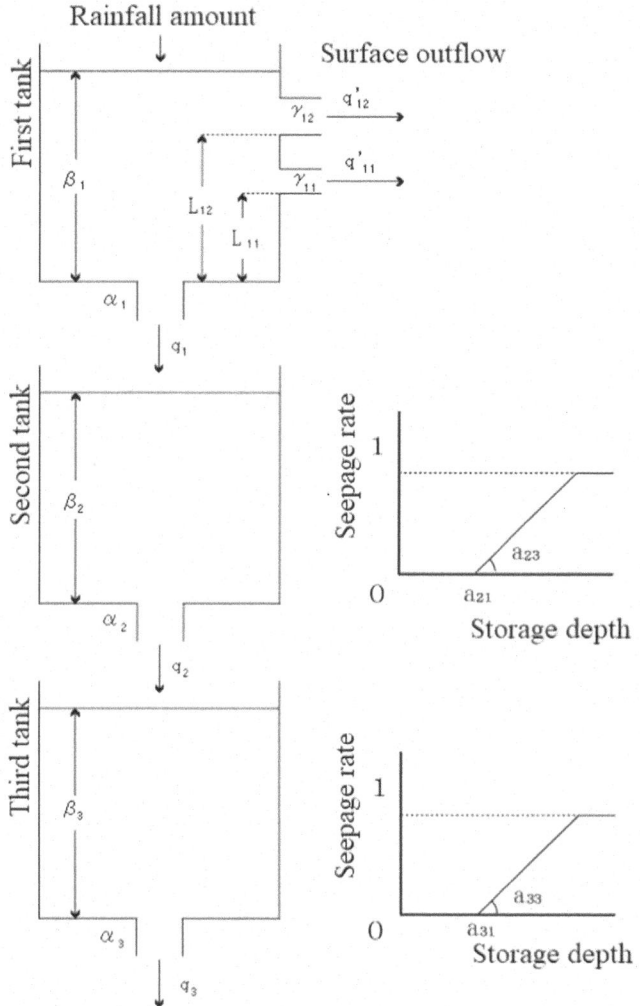

Figure 8.1 Structure of the Aoki-Hioki model.

In 2014, Totsukawa Village in Nara Prefecture signed a partnership agreement with the Osaka Institute of Technology and further strengthened the "monitoring activities of deep-seated landslide risks during torrential rains." The Laboratory for Geotechnical Disaster Prevention of the same institute installed unique rain gauges in three locations within Totsukawa Village in 2015 (Asahi, Imoze, and Detani, Photo 8.1) to automate the analysis process (calculation of the LRI). Since 2016, rainfall data and

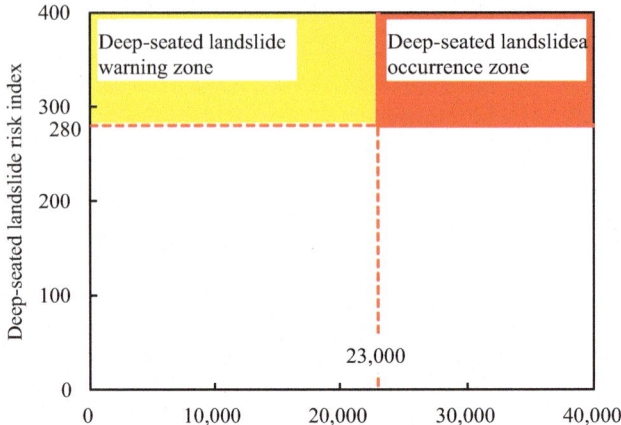

Figure 8.2 Index for the risk of deep-seated landslides caused by torrential rains in the southern part of Nara Prefecture.

Photo 8.1 Rain gauge installed in Totsukawa Village by the Laboratory for Geotechnical Disaster Prevention (Photo by Kazuaki Hioki).

analytical results (the LRI and its one-week cumulative value) of the three locations (Asahi, Imoze, and Detani) are automatically sent to the village and the institute every 60 minutes.

At present, the laboratory monitors the risk of deep-seated landslides during torrential rains at 10 locations in Totsukawa Village (Figure 8.3). An example of the monitoring results is shown in Figure 8.4. Since the beginning of this monitoring, analytical results have never reached the red zone, and the yellow zone was only reached twice (Typhoon Nangka in 2015 and Typhoon Krosa in 2019). The Totsukawa Village Hall fully understands the difference

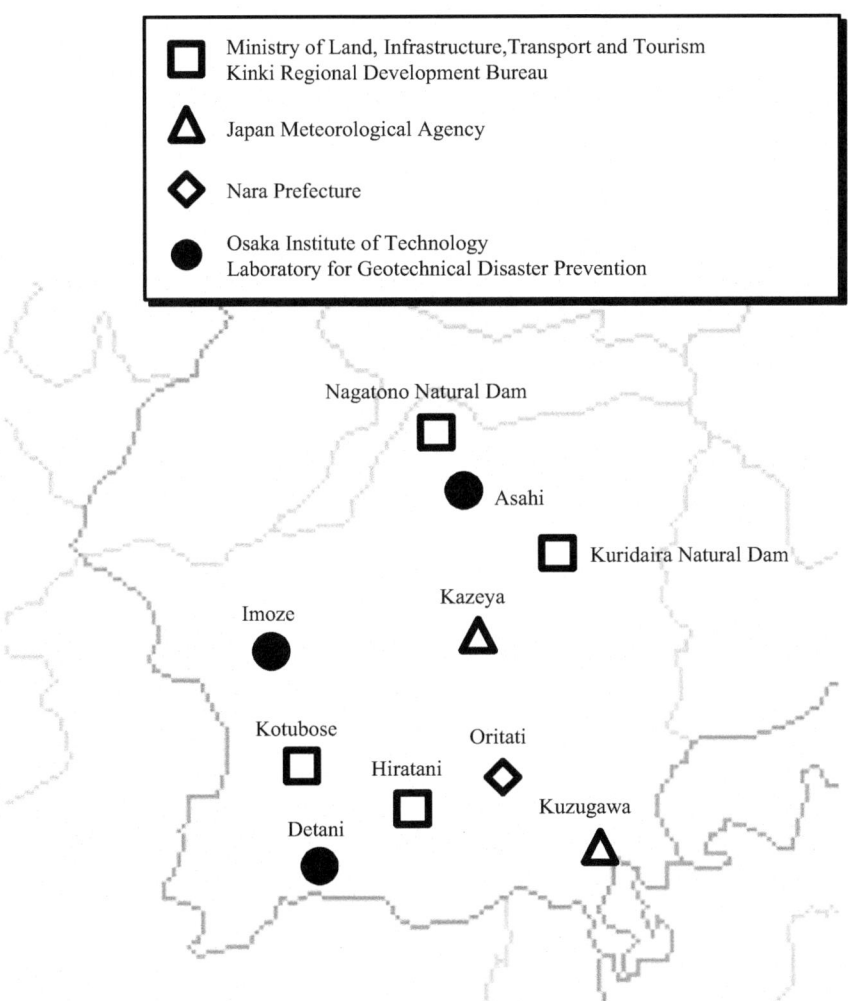

Figure 8.3 Monitoring points for deep-seated landslides during torrential rains at 10 locations in Totsukawa Village.

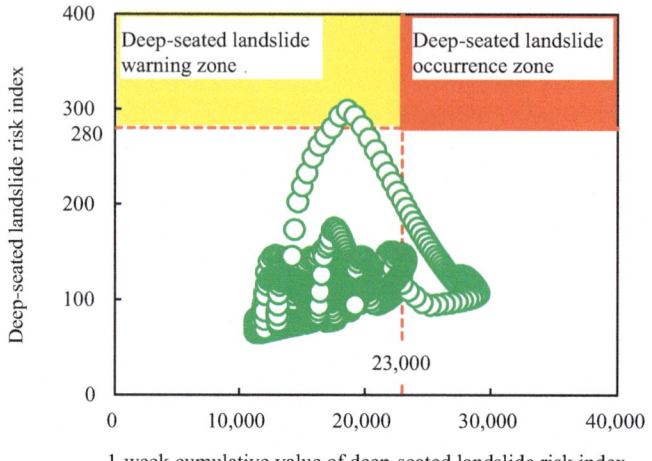

Figure 8.4 An example of the monitoring results (April to October 2019 Detani).

between the landslides warning (jointly issued by Nara Prefecture and the Japan Meteorological Agency) and the deep-seated landslide warning (issued by the laboratory) and actively uses both warnings. In particular, the latter warning is useful in making decisions regarding the evacuation of residents. The laboratory plans to continue the monitoring of deep-seated landslide risk during torrential rains in the southern part of Nara Prefecture around Totsukawa Village with the hope that the monitoring is useful in enhancing disaster-prevention and mitigation ability in this area.

8.2.2   Flood and Inundation Measures Taken by Shingu City and Other Local Governments

In the southern part of Kii Peninsula, flood and inundation damage occurred over a wide area. For this reason, local governments in southern Wakayama Prefecture implemented new flood and inundation measures as shown in Table 8.1.

8.2.2.1   Non-structural Measures Based on Local Characteristics

The manner in which flooding and inundation occurred revealed that different factors were at play in each affected area. Therefore, local governments revised their disaster-prevention manuals and reviewed their operations based on the disaster situation in each area.

In Shingu City, whose administrative district encompasses the middle/lower reaches of the Kumano River to its mouth, river flooding occurred differently in the urban area located at the mouth of the river and in the mountainous area around the Kumano River basin.

Table 8.1 Flood and inundation measures after the Kii Peninsula Disaster

Non-structural Measures	Structural Measures
1 Revision of the disaster-prevention manual and review of operations a Revised the manual to account for changes in river levels and rainfall b Established high-priority flood control points c Established warning zones where mandatory measures can be taken d Clarified the evacuation support for those in need of assistance 2. Improvement of the administrative staff's disaster preparedness a To double as training, staff are instructed to begin disaster-prevention activities from the moment evacuation preparation information is announced b Conducting staff training on the Kii Peninsula Disaster	1 Reinforcement of the information-collection system a Created a system for river monitoring 2. Disaster mitigation measures in the event of flooding a Established evacuation routes to the hills in back of residential areas b Established park-like areas on high ground for evacuation 3. Disaster mitigation measures to ensure rapid reconstruction a Established lots on high ground for evacuation of residents' vehicles b Relocated power supply facilities for the water supply, etc., to higher locations c Relocated high-voltage power receiving equipment for buildings to higher locations d Relocated emergency power generators to higher locations e Relocated outdoor air conditioner units to higher locations
3. Improvement of residents' disaster preparedness a Holding disaster-prevention training for residents	4. Development of disaster-prevention bases a Established facilities that also serve as disaster- prevention bases on developed hills

The urban area had a history of frequent flooding from inadequate drainage. To prevent such flooding, sluice gates and pumping stations were constructed on two small rivers that flow through the area, and these measures proved to be effective. However, during the Kii Peninsula Disaster, torrential rain that exceeded the capacity of these facilities caused flooding from inadequate drainage that lasted from late night on September 3 to early morning on September 4. Additionally, following this, overflows occurred at three points on the levees, and muddy floodwater from the main channel of the Kumano River flowed into the urban area River flowed into the urban area. Based on how this disaster occurred, the city revised its disaster-prevention manual to establish flood warning zones (zones where mandatory evacuation measures could be taken) in the areas where water levels had risen significantly.

In the mountainous area around the Kumano River Basin, the Kumano River and its tributaries began flooding on the evening of September 2, causing inundation above floor level across a wide area. The rise in water level was particularly high in the areas around and upstream of the Kumano River Branch Office. The floodwater reached the second floor of the Kumano River Branch Office on the early morning of September 4. This pattern of flooding was regulated by the topographical characteristics of the Kumano River. In other words, there is an 18-kilometer-long section where the river narrows in the lower reaches of the Kumano River. This caused a significant rise in water level in the areas upstream of the narrow section of the river. Based on how this disaster occurred, the city revised its disaster-prevention manual to establish high-priority flood control points in the areas surrounding the Kumano River Branch Office.

In Nachikatsuura Town, home to one of the three Grand Shrines of Kumano (Kumano Sanzan) that have been listed as UNESCO World Heritage sites, conditions suddenly became catastrophic along the Nachi River from Nachi-no-Otaki Falls, the symbol of this sacred site. Early in the morning of September 4, torrential rain caused large amounts of water and debris to flow into the Nachi River from all of its tributary streams. This in turn caused the river to overflow and the water level to rise so rapidly that floodwater reached the eaves of houses. The headwaters of the Nachi River are located in steep mountains that rise 800 to 900 meters above sea level, and the river's rapid flow is only about seven kilometers long, reaching the sea via Nachi-no-Otaki Falls. Because of these river characteristics, fluctuations in rainfall were quickly and clearly reflected in changes in the river level during the Kii Peninsula Disaster. Based on how this disaster occurred, the town revised its disaster-prevention manual to account for information on river water levels as well as the amount of rainfall in the river basin.

Although no human casualties occurred in Kozagawa Town, where there is an aging population, the town reviewed operations according to its disaster-prevention manual and prioritized evacuation support activities for those in need of assistance, in order to further improve such activities. In addition, since many staff members retired after the flood disaster, resulting in an increased percentage of administrative staff without experience in disaster response, the town is aiming to improve disaster preparedness by starting disaster-prevention activities sooner than scheduled.

8.2.2.2   *Structural Measures to Speed Up Reconstruction*

During the disaster, electrical facilities that support the infrastructure for daily life suffered flooding and broke down, resulting in prolonged water and power outages that became a major obstacle to restoration activities. In particular, it became clear that the flooding and breakdown of electrical equipment directly led to water outages.

In Shingu City and Nachikatsuura Town, water intake pumps and electrical equipment for the water supply became flooded and broke down, and it took several days to procure parts to repair them, which significantly delayed the restoration of the water supply. In Kozagawa Town, a building with a water supply tank on its roof was unable to operate the water pump, resulting in a prolonged water outage. In addition, high-voltage power receiving equipment that delivers power to buildings got flooded and broke down, meaning that some buildings experienced delays in restoring power after the blackouts, even after the power transmission and distribution company resumed power transmission. Based on how this disaster occurred, some local governments, hospitals, and elderly care facilities have relocated electrical equipment for the water supply, high-voltage power receiving equipment and emergency power generators for buildings, and outdoor air conditioner units to higher locations as a part of restoration work or facility upgrades.

Because many vehicles were inundated when the river flooded, there was a shortage of vehicles (especially mini trucks) that could be used for cleanup during restoration. In light of this, some local governments have constructed lots on high ground where residents' vehicles can be evacuated ahead of a disaster. In the Nachi area of Nachikatsuura Town, where many people fell victim to the disaster, measures were taken to construct a new evacuation route for residents in the hills behind the residential district.

8.2.3 Development of Kiho Town's Version of a Timeline

In Kiho Town, Minamimuro County, Mie Prefecture, the Kii Peninsula Flood caused road closures and flooded houses throughout the town. At the peak of the precipitation, evacuation orders were issued for 7,432 people in 3,405 households, and approximately 1,000 people, including those who voluntarily evacuated, went to 19 evacuation centers in the town. Although the town had completed the construction of a ring levee, raised roads and houses, and installed drainage pump stations as hardware countermeasures, the disaster was a major reminder of the limitations of hardware counter-measures against unforeseen external forces.

8.2.3.1 Effectiveness and Issues of Self-Help and Cooperation

Figure 8.5 shows the evacuation of residents in a certain district of Mie Prefecture that was inundated by the Kii Peninsula Flood. Of the 451 residents who received evacuation orders, 69.8% took evacuation action, including going to shelters and other locations, and 23.5% of all residents did not evacuate despite being inundated. Only 24.6% of the residents evacuated to designated evacuation sites, and many evacuated to other places, such as outside of the village or to houses that were not flooded. Furthermore, it was found that only 7.1% of the residents voluntarily evacuated to evacuation sites, and 17.5% were rescued by boats after the flooding. This means that more than half of those who went to evacuation sites were rescued and evacuated.

Interviews were conducted with those who did not evacuate by themselves and those who were rescued. They said, "I did not think that the water would flow over the ring levee," "The water had never come this far even in Isewan Typhoon," and "Even if we were flooded up to the second floor, we could just

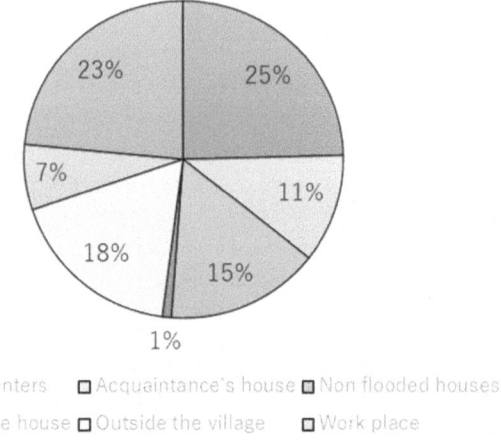

Figure 8.5 The evacuation condition of residents in an inundated district.

swim across to the hill behind my house." The Isewan Typhoon, internationally named Typhoon Vera, caused extensive damage in modern Japan. It occurred in 1959 and claimed more than 5,000 victims. Overconfidence in hardware measures and self-confidence based on past experience may have contributed to the low evacuation rate. Although hardware measures are required, it is also necessary to educate people to evacuate voluntarily when an evacuation advisory and evacuation order is issued, without overconfidence in hardware measures.

Although self-sufficiency is the basic principle for evacuation during a disaster, many rescue activities (cooperate activities) led by the town employees, police, fire department, neighborhood residents, and voluntary disaster-prevention organizations were conducted and proved effective. On the other hand, the residents who actually participated in the rescue activities reported that they did so with a sense of fear that a secondary disaster might occur. It is necessary for disaster-prevention organizations, including voluntary disaster-prevention organizations, to establish objective criteria and share information regarding the continuation or discontinuation of rescue activities.

8.2.3.2   The Kiho Town Timeline in Mie Prefecture

An advance disaster-prevention action plan (timeline) is a new concept in disaster-prevention response. The timeline is a method of minimizing damage by organizing "when," "who," and "what to do" for foreseeable disasters along a time axis and formulating these plans in advance among the parties concerned. In the case of typhoon-induced wind and flood damage, it is important to decide in advance on actions to be taken. In the case of Major Hurricane Sandy that hit the United States (2012), it was reported that prompt disaster management actions by disaster management organizations according to a timeline that had been prepared in advance led to significant disaster-mitigation effects.

In February 2014, Kiho Town, Mie Prefecture, became a pioneer in Japan by establishing a Timeline Formulation Study Group and has been working to formulate the Kiho Town Timeline, which incorporates concepts from the United States. Nineteen organizations, including the town, national government, the prefectural government, fire department, Kiho Town–related organizations, and private companies, participated in the discussions to formulate the plan. The expected patterns of action numbered approximately 220 items, and the "what to do" and "when to do it" in the event of a disaster were arranged by the target organizations. As shown in Table 8.2, in the Kiho Town Timeline, the time of the typhoon's closest approach or landfall is set as the base time (0 hour), and all public organizations are to begin preparations 120 hours (5 days) in advance, with each organization completing actions from patrols to evacuation by 0 hour. During the trial phase, some of the comments from those involved included, "No more omissions in disaster response," "Awareness of all organizations involved can be shared," and "Disaster prevention actions can be taken with a little more time to spare."

On February 21, 2015, Kiho Town, the Kinan Office of Rivers and National Highways of the Kinki Regional Development Bureau, the Kisei National Highway Office of the Chubu Regional Development Bureau, and the Tsu District Meteorological Observatory concluded an Agreement on the Collaboration of an Advance Disaster-Prevention Action Plan (Timeline) for

Table 8.2 The layout of the Kiho Town Timeline

When	Who	What to Do
120–0 hours before	All public organizations	Collection and sharing typhoon and other weather information
96–72 hours before	Towns, governments, prefectures, etc.	Reconfirmation of prior disaster-prevention actions within the municipality
96–72 hours before	Towns, governments, etc.	Patrol and inspection of town facilities
72–48 hours before	Towns, governments, prefectures, etc.	Assumption and adjustment of evacuation plans
72–36 hours before	Towns	Advance notification to residents and warning of disaster-prevention actions
48–36 hours before	Towns	Establishment and coordination of evacuation centers
36–12 hours before	Towns, firefighters, police, etc.	Call for evacuation and guidance
12–6 hours before	Towns, firefighters, police, etc.	Rescue and evacuation guidance
0–72 hours later	All public organizations	Rescue, salvage, and recovery

Typhoon and Other Windstorm Disasters in Kiho Town, which was the first such agreement in Japan. Although the formulation of the plan required detailed prior consultation among many related organizations, it is expected to be highly effective in mitigating disasters.

8.3   Initiatives by Road and Railway Operators

8.3.1   *Highway Slope Health Diagnosis System during Heavy Rainfall*

Rainfall-induced slope failures have become more frequent owing to recent occurrences of extreme weather in Japan. In particular, the early detection of a sudden rainfall-induced slope failure is known to be relatively difficult compared with the detection of typical slope failure behavior. In response to this problem, Japanese-expressway-operating companies have their own regulation standards to prevent rainfall-induced slope disasters based on historical rainfall data. However, sometimes the judgment of whether a slope failure will occur is not accurate because rainfall information does not directly reflect the soil-moisture condition of a slope.

To solve this problem, Osaka University and West Nippon Expressway Company Ltd. (NEXCO-West) have proposed soil moisture–based indexes for the structural-health monitoring of a slope against a rainfall-induced slope failure under newron® Project. Figure 8.6 shows a method for evaluating the health condition of a slope by monitoring soil moisture–infiltration behavior. The relationship between volumetric water content and displacement (shear deformation) to slope failure shown in Figure 8.6 is based on previous work [4].

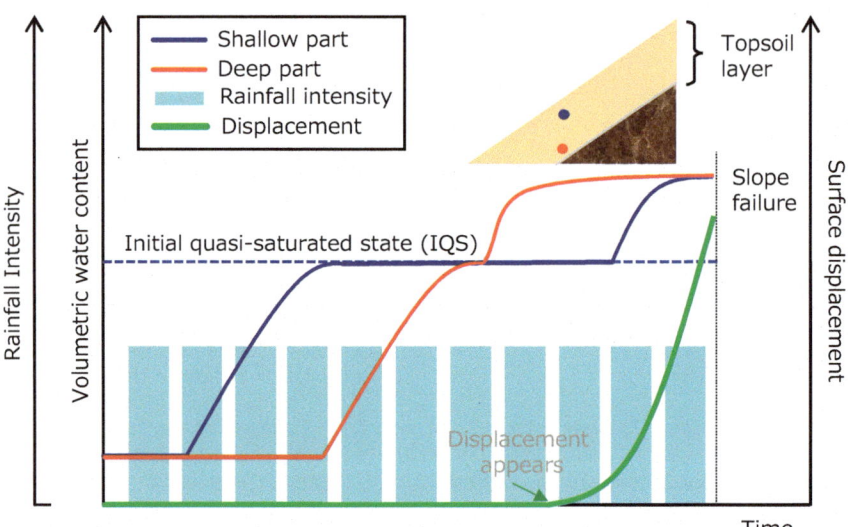

Figure 8.6 Schematic diagram showing the relationship between volumetric water content and displacement to slope failure due to rainfall.

When rainwater infiltrates the model slope, the volumetric water content at its shallow part (blue point) starts to increase first, and then the volumetric water content marked in red, above the boundary part, starts to increase and reach θ_{IQS}, respectively. Here, θ_{IQS} indicates the temporary equilibrium state of volumetric water content due to the balance between rainfall intensity and soil permeability under constant rainfall intensity conditions. This shows that the entrapped air remains in the soil void, and this phenomenon is called the quasi-saturation phenomenon. Subsequently, if the rainwater continuously infiltrates into the slope, the groundwater layer forms at the bottom of the surface layer, and the volumetric water contents start increasing again from the boundary part. This implies that the entrapped air in the soil void replaces the pore water. And the volumetric water content reaches θ_{FS} which is the field-saturated volumetric water content, sequentially from the deep part. Then, a slope failure is caused by the decrease in effective stress owing to water pressure in the pores. The risk of slope deformation increases after the volumetric water content in the shallow part goes through θ_{IQS} and the volumetric water content in the deep part reaches θ_{FS}. Therefore, by observing θ_{IQS} and θ_{FS}, the rainwater infiltration behavior of the slope can be determined. In addition, by combining θ_{IQS} and θ_{FS} with the Internet of Things (IoT), the health of slopes during rainfall can be diagnosed in real time. Here, a case study of this method in an actual slope using LPWA (Low-Power Wide Area), which is one of the IoT technologies will be presented.

LPWA is one of the wireless communication standards that uses the sub-GHz band with a low-power and wide-area network. Figure 8.7 shows an overview of the installation of each sensor and LPWA instrument (L-Watch, EW) on the slope along the expressway as part of the NEXCO-West Real-time Observation Network (newron®). The geology of the slope is weathered granite, and the topsoil layer is decomposed granite soil. Here, the boundary layer between the topsoil layer and the base layer was set as $N_d = 12$ using the results of a simple dynamic cone penetration test. The boundary layer is usually determined by $N_d = 10$, but here it was determined by considering the test results. Soil moisture sensors (SM-150T and Delta-T) were installed at depths of 0.3 m and 0.8 m in the middle and toe of the slope, respectively, to observe quasi-saturated and field-saturated phenomena in shallow and deep parts. A tilt sensor to detect slope movement is built into the L-Watch. A rain gauge (RS-102N, ANEOS) is installed on the same slope. Data acquired from each sensor are collected every 15 minutes by the L-Watch and can be observed on the web screen (C-Watch, EW) via an existing base station (SIGFOX, sigfox). Unlike conventional wireless observation systems, there is no need to set up a base station, reducing the time and cost required for installation and relocation.

Here, the θ_{IQS} shown in Figure 8.6 is positively correlated with rainfall intensity, so it is necessary to obtain the θ_{IQS} (hereafter referred to as IQS index) for the rainfall intensity that changes from time to time to apply the

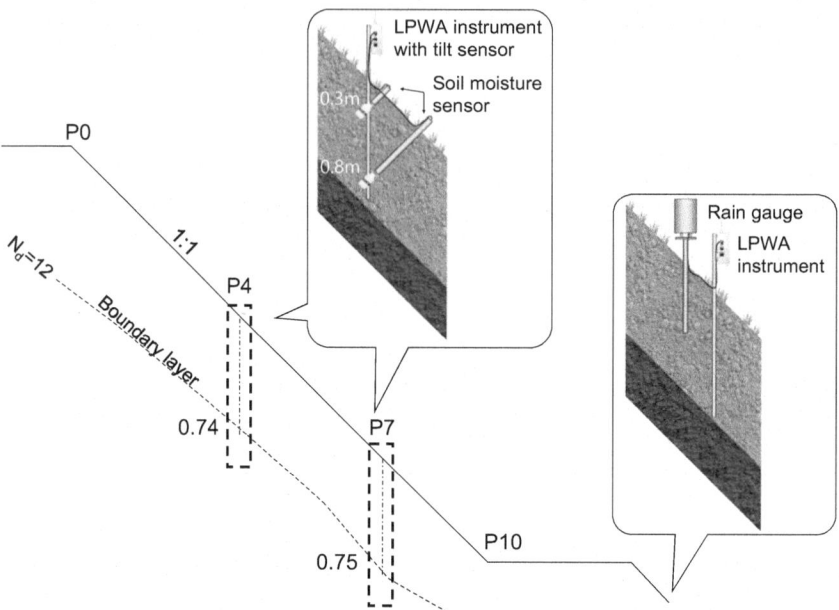

Figure 8.7 Installation overview of sensing instruments.

method to actual slopes. See Koizumi et al.'s [4] for the method to obtain the IQS index. θ_{FS} is then obtained from an undisturbed soil sample at a depth of 0.8 m. The FS index is calculated by adding degassed water to the sample and determining the volumetric water content when the sample is saturated. Note that 1.0 for both the IQS and FS indexes indicates quasi-saturation and field saturation, respectively.

Figure 8.8 shows an example of a web observation screen that visualizes the slope condition using the sensing data of the toe of the slope collected by L-Watch. The right side of the screen shows the current IQS index, FS index, and value of the tilt sensor and their status with face marks. This information provides an accurate diagnosis of the current state of rainwater infiltration on the slope. Next, the left side of the screen shows the time series variation of each data for the past week. The top shows the variation in the tilt sensor, the second shows the variation in the IQS and FS indexes, the third shows the variation in volumetric water content at a depth of 0.3 m and 0.8 m, and the bottom shows the 15-minute rainfall intensity. From this information, it can be confirmed that no deformation has occurred on the slope and that the IQS and FS indexes are below the standard value of 1.0. From the above information, it is possible to visualize the behavior inside the slope from unsaturation to saturation and deformation just by checking this screen.

Figure 8.8 Visualization of slope health during rainfall.

Management of slopes to prevent rainfall-induced slope failure has become an important issue in Japan with the recent increase in extreme weather conditions. To solve this problem, we propose IQS and FS, soil moisture–based indexes, and introduce the result of a case study on an actual slope along the expressway in Japan. The research findings are as follows. A real-time monitoring system for slope failure was developed utilizing IoT to function under the NEXCO-West Expressway Real-Time-Observation Network, newron®. By visualizing the slope condition using the IQS and FS indexes, it was shown that slope health can be determined from the rainwater infiltration behavior in the slope, which is difficult to determine only by observing the rainfall information.

8.3.2 Examination of New Standards for Responses to Heavy Rainfalls on Train Operations

On railways, standards for regulating heavy rainfalls are laid down for railway lines to ensure safety by patrol, to slow down or stop trains, in sections where a rain hazard might occur. However, in recent years, when brief heavy rains or record-breaking rains are increasing, regulating a response to heavy rainfalls has struggled to correspond to the changes. With this background, West Japan Railway Company (hereafter, JR West) introduced new standards for responding to heavy rainfalls, analyzing the precipitation during the past disasters and the precipitation recurrence interval, while promoting rain-hazard reduction measures. In addition, we studied standards for train operations to respond to heavy rainfalls on the train operations of JR West railway lines, using the precipitation data from radar by external weather information (hereafter, radar precipitation); here, we introduce the summary.

8.3.2.1   The Summary of Standards for Regulating Responses to Heavy Rainfalls on JR West Train Operations

The standards for regulating responses to heavy rainfalls on the train operations of JR West are shown in Table 8.3. When these values exceed these standards, JR West regulates the train operations, or patrol along railway lines. If the five-day cumulative precipitations exceed these standards, JR West lowers the standards ①→② in Figure 8.9.

In the past, when the values of continuous precipitation were not reset until the rain lasted for more than 12 hours, there were the following problems.

If the rain stopped and fell repeatedly, the measurement of continuous precipitation is continued. For this reason, if the continuous precipitation exceeded these standards, the train operations couldn't restart, and the oversight of train operations continued for a long time. On the other hand, if the rain stopped for more than 12 hours, the continuous precipitation was reset for the moment, even if the rain continued intermittently for long durations. So, in

Table 8.3 The precipitation to use for the regulating responses to heavy rainfalls on train operations

Precipitation	Summary
Hourly precipitation	Amount of rainfall from 1 hour ago
Continuous precipitation	Amount of rainfall from 24 hours ago
Cumulative precipitation	Amount of uninterrupted rainfall from 48 hours ago
5-day cumulative precipitation	Amount of rainfall from 5 days (120 hours) ago

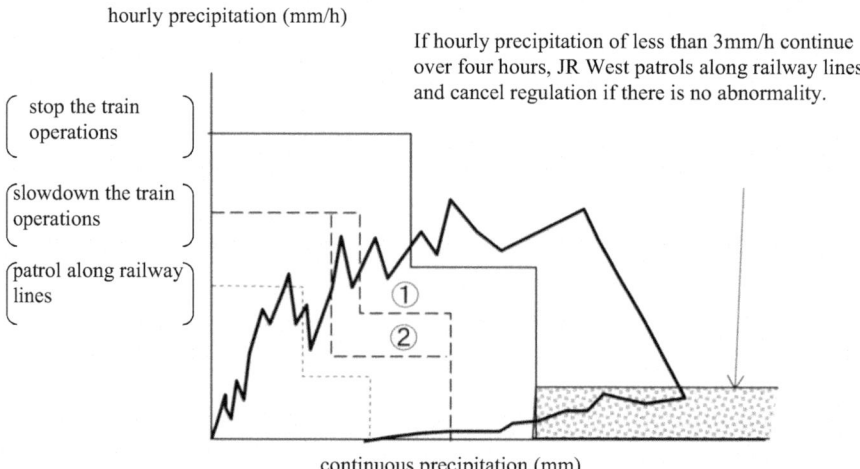

Figure 8.9 Conceptual diagram of the standards for regulating heavy rainfalls.

some cases, the influence of previous rainfalls was not reflected in the continuous precipitation numbers. For problems like this, in June 2015, JR West modified the rules regarding continuous precipitation to be the amount of rainfall measured in 24 hours. This modification made it possible to respond to the changes of the rain condition from moment to moment. Normally, when train operations are stopped because of heavy rain, the maintenance staff patrol railway lines and, if there is no abnormality, restart the train operation. On the other hand, even if the continuous precipitation exceeds the regulation value, train operation can be resumed by confirming that no abnormality exists on railway lines, if the hourly amount of rainfall is less than 3 mm/h and continued more than four hours. This standard was applied based on the result of the experiment on the model embankment with a water sprinkling.

8.3.2.2 Setting the Standards for Regulating the Response to Heavy Rainfalls on the JR West Train Operations

We analyzed the precipitation of the past disasters by rainfall that occurred in the area operated by JR West. Here, we introduce the results of precipitation analysis along railway lines where disasters often occurred.

Figure 8.10 shows the frequency distribution of hourly precipitation that was analyzed along railway lines when rainfall disasters occurred in the past. According to Figure 8.10, disasters occurred frequently at the hourly precipitation rate of 10~30 mm/h, and precipitation of less than 30 mm/h accounted for 70%. On these railway lines, the average hourly precipitation at the time of the rainfall disasters was 27 mm/h, and the standard deviation was 14 mm/h. Then, using these data, we analyzed the relationship between the

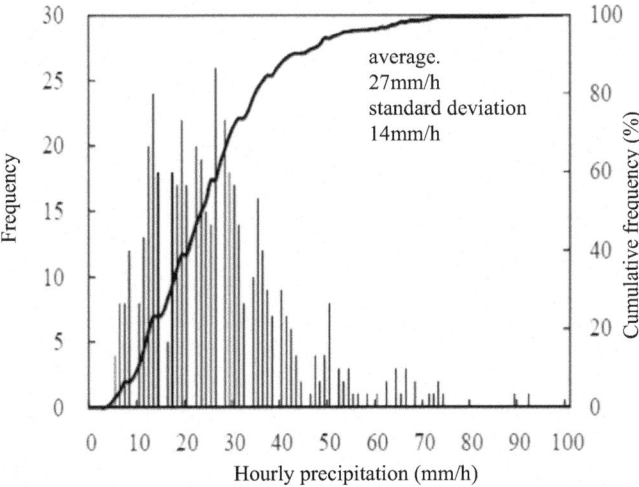

Figure 8.10 Frequency distribution of hourly precipitation when disasters occurred.

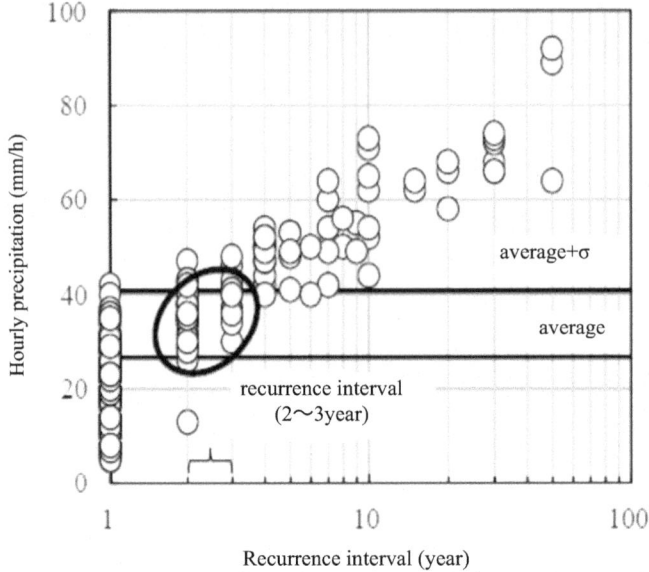

Figure 8.11 Recurrence interval of hourly precipitation when disasters occurred.

hourly precipitation and the precipitation recurrence interval for the past rainfall disasters (cf. Figure 8.11). We calculated the recurrence interval of the precipitation from the Gumbel distribution, using the data from the AMeDAS (Automated Meteorological Data Acquisition System) observatory nearest the disaster occurrence sites. As a result of this analysis, we found out that the hourly average precipitation at the time of the rainfall disasters was 27 mm/h, and the recurrence interval was a period of 2–3 years. We analyzed a similar relationship for continuous precipitation; the precipitation recurrence interval was about the same as for hourly precipitation.

We analyzed a similar relationship of hourly and continuous precipitation recurrence interval on the railway lines of JR West, which have different resistances to rain hazards. As a result of this analysis, JR West determined the value at which to slow down train operations on the target railway lines using average rainfalls at the time of the past rainfall disasters. In addition, JR West determined the value at which to stop train operations on the specific railway lines using an average rainfalls+1σ (σ: standard deviation). On the other hand, JR West lowered the value on the railway lines where the possibility of rain hazards was low because of its rain-hazards reduction structures.

8.3.2.3   *Meteorological Hazard Management System*

On JR West railway lines, train operations are regulated in the way mentioned above, by using the precipitation observed by the rain gauges (average

13 km interval) installed along the railway lines. On the other hand, with the advancement of meteorological radar systems, it has become possible to observe local heavy rainfalls in recent years. On the railways, it is necessary to use informations like this for safe train operations.

For this reason, JR West introduced the meteorological hazard-management system on the main railway lines in the area around Kyoto, Osaka, Kobe in 2018. This meteorological hazard-management system can unitarily manage the regulations of these railway lines.

This meteorological hazard-management system has the following features:

• This system can handle local heavy rainfalls using the radar precipitation per 1 km and the rain gauges installed along the railway lines, and it can distribute this information quickly.
• This system can also display the information of the sections and phenomena requiring regulation by strong wind, earthquake, and so on.

JR West gradually introduced the system of only radar precipitation on all the remaining railway lines. We show the image of the meteorological hazard-management system in Figure 8.12.

We are promoting the rain hazards reduction measures in JR West railway lines. We are also promoting safety by regulating the train operations in heavy rainfalls when a rain hazard may occur. Even more, we would like to use this system to capture local heavy rains and further improve the safety of train operations.

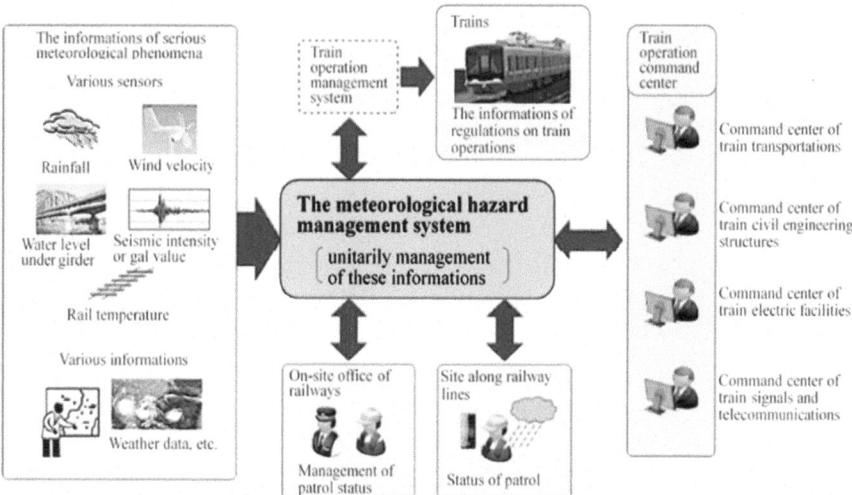

Figure 8.12 The image of the meteorological hazard-management system.

8.4   Summary

In this chapter, we introduced the disaster-prevention and mitigation measures that were implemented after the Kii Peninsula Disaster. In severely affected areas, each municipality took the initiative to enforce disaster-prevention and mitigation measures through, for example, the introduction of timelines, by making revisions to disaster-prevention manuals, and by implementing hardware measures. In addition, various monitoring systems were developed and implemented using observation, measurement, and communication devices. Under frequent localized and sudden torrential rains, the need is increasing for precise and real-time monitoring technology that can incorporate geological characteristics and rainfall information from each location. We hope that these measures and their successes are shared between communities and organizations to offer systematic and synergistic improvements in the disaster-prevention and mitigation ability of sediment and flood disaster-prone countries.

References

[1] Okada, K.: Soil water index, *Weather Service Bulletin*, Vol. 69, No. 5, pp. 67–100, 2002. (in Japanese).

[2] Nakamura, S. and Hioki, K.: Parameters of three-tier tank model used for forecasting sediment-related disaster during heavy rainfall, *Journal of the Japan Landslide Society*, Vol. 60, No. 1, pp. 1–15, 2023. (in Japanese).

[3] Hioki, K., Saratani, Y., Kamatsuka, Y., Nakamura, S., Kojima, T. and Fujita, T.: Hazard indicator and application to predict downpour-induced deep-seated landslides in southern Nara, *Memoirs of Osaka Institute of Technology*, Vol. 63, No. 1, pp. 15–27, 2018. (in Japanese).

[4] Koizumi, K., Komatsu, M., Oda, K., Ito, S. and Tsutsumi, H.: A consideration for evaluating of slope health monitoring during rainfall focusing on volumetric water content, *Journal of Japan Society of Civil Engineers, Series C*, Vol.77, No.2, pp. 129–139, 2021. (in Japanese)

9 Response to Sediment and Flood Disasters Caused by "Unexpected" Heavy Rainfall and Lessons Learned

Ryoichi Fukagawa

9.1 Introduction

In 2011, record-breaking torrential rains caused by Severe Tropical Storm Talas caused extensive sediment disasters and flood damage mainly in the Kii Peninsula. The Research Committee on Sediment Disasters Caused by "Unexpected" Heavy Rainfall investigated the damage situation and disaster-generation mechanism in the disaster area of the Kii Peninsula and obtained many findings through interviews with residents of the disaster area and support activities. Based on the results of a series of committee activities, this chapter summarizes the responses to and lessons learned from sediment and flood disasters caused by heavy rainfall in the form of recommendations for governments, facility managers, and the general public.

9.2 Consider the "Unexpected"

Six months before the Kii Peninsula Disasters, the word "unexpected" was thrown around in the mass media in the wake of the catastrophe caused by the Tohoku-Pacific Ocean Earthquake. The word "unforeseen" had a variety of meanings and was interpreted in many different ways by many people. It is used to mean "unexpected," "beyond design conditions," "extremely high level of damage," "disaster of an unpreventable scale," among others, and some news media interpreted it as an evasion of responsibility by facility managers, researchers, and engineers. Although the word "unexpected" is innocent, the impression of the word has become worse, and there was even hesitation to use the word. The term "unexpected" was used in the name of the Committee. This is because we bear in mind that when a disaster occurs, there is a possibility that the level of external forces may exceed the level used in the design process, etc., and also because it is a reminder that we, as researchers and engineers, must take disaster seriously.

Structures are designed to withstand external forces of a certain magnitude. For disaster-prevention facilities, external forces that can withstand the largest disaster in the past are often set. This level of external force can be called "within expectation," which is the concept of "hard measures"

DOI: 10.1201/9781003375210-9

whose main purpose is to prevent disasters from occurring. For disasters of a scale that cannot be prevented by hardware measures, various soft measures are required to prevent the spread of damage. Even in the event of a disaster that exceeds the design forces, it is necessary to assume the scale of damage and implement countermeasures to prevent the occurrence of "unexpected" disasters. This is a realistic approach to "unexpected" disasters.

As researchers and engineers, when considering disaster prevention and mitigation, we are required to keep the meaning of "unexpected" fully in mind and strive to conduct research and work on specific measures and forecasting methods to deal with unexpected disasters from ordinary times.

9.3 Recommendations on Disaster Forms

9.3.1 Sediment Disasters

The sediment disasters that occurred in the Kii Peninsula varied in form according to topography and geology, with large-scale slope failures occurring in accretionary bodies and large-scale debris flows occurring mostly in igneous bodies. In addition to the fact that large-scale slope failures occurred when cumulative rainfall increased and surface failures and debris flows occurred when short-time rainfall increased after cumulative rainfall increased, it was found that large-scale slope failures and debris flows have repeatedly occurred under similar conditions in the past.

Recommendations to the administration and facility managers resulting from these findings are as follows.

9.3.1.1 Know the Relationship among Rainfall Patterns, Geologic Conditions, and Sediment Disasters

In the Kii Mountains, when the cumulative rainfall exceeds 600 mm, large-scale slope failures are likely to occur, especially in accretionary geological formations. On the other hand, in the igneous geology, after the cumulative rainfall reaches 700~800 mm, surface failures and debris flows occur frequently due to the hourly rainfall of 70~80 mm or more. Since the relationship between sediment disasters and rainfall characteristics is greatly influenced by geological bodies, it is desirable to actively utilize the JMA method of soil rainfall indexes and hydrological condition indexes within slopes using tank models, based on an understanding of geological bodies. In situations where the soil rainfall index exceeds the warning level, it is necessary to take sufficient precautionary measures in cooperation with the local community.

9.3.1.2 Predict Sediment Disasters

If there are landslide landforms (large-scale slope failure sites) in the vicinity of the affected area, caution should be exercised. In addition, there is a possibility of flood disasters occurring in the vicinity of the affected area because of river channel blockage in the upper reaches of rivers. On the other hand, surface failures and debris flows are more likely to occur when short-term rainfall increases after the cumulative rainfall increases. The locations of surface failures and debris flows should be determined by referring to land condition maps and slope failure warning zones established by pre-fectural governments in areas where there are houses and other preserva-tion targets, as well as identifying hazardous areas based on local residents' experiences and traditions. It is also desirable to develop hazard maps of sediment disasters that enable residents to visualize the damage. In addi-tion, it is desirable for the government and facility managers to make effective use of these hazard maps to educate local residents and promote cooperation.

9.3.1.3 Build Partnerships for Sediment Disaster Mitigation

In the event of a surface failure, debris flow, or large-scale slope failure, it is necessary to study how to establish a sufficient cooperative system among related departments in anticipation of a sediment disaster to enable early res-toration. In addition, it is desirable to establish a cooperative system between the government, facility managers, and private companies such as construction-related industries in order to enhance emergency response capabilities in the event of a disaster.

9.3.1.4 Identify the Disaster Resilience of the Target Site and Utilize a Monitoring System

It is important to identify areas where large-scale sediment disasters are likely to occur based on topography, geology, and disaster history, and to develop a system to monitor large-scale sediment disasters using the latest monitoring and sensing technologies and so on, so that local residents can take eva-cuation actions.

9.3.2 Flood Disasters

River maintenance is planned and built to cope with external forces of a certain magnitude. In the recent torrential rains, in areas where external forces were generated in excess of the planned external force, phenomena such as damage from the levee overflowing and external forces acting from the inside of the levee when floodwaters entering the inner levee area returned to the outside of the levee were observed. In some areas, bridges, embank-ments, and other structures around the river that were constructed based for a specific amount of water were damaged by flooding when they received rainfall exceeding the probability of the plan. The extent of inundation was generally consistent with that determined from the topography and flood history around the river.

The followings are recommendations to the government and facility man-agers based on the above findings.

9.3.2.1 Learn from Past Flood Traces and History

Since there is a limit to the scale of flooding that can be controlled by river improvement, it is essential to actively disseminate information to residents on the "possibility of flooding up to this point" based on past flood traces and flood history, as well as the history of land formation where residents reside, to inform residents of the risks of flooding in their land and region. Photo 1.1 shows the monuments indicating the maximum water levels of the Meiji, Showa, and Heisei Kii Peninsula Disasters. It is hoped that the lessons of these disasters will be passed on for 100 to 200 years.

9.3.2.2 Pass on Local Flood Lore to Future Generations

Residents in areas that have experienced flooding have wisdom on how to mitigate risks associated with flood characteristics unique to the area. It goes without saying that the government should make use of this information, but it is also necessary for the residents themselves to pass on this knowledge to strengthen the disaster preparedness of the entire community.

9.3.2.3 Recognize the Possibility That "Unexpected" Events May Occur

River development is planned and constructed for external forces of a certain magnitude, and it may be impossible to completely control floods when external forces exceeding the design occur. If local residents understand this fact, the government and facility managers will be able to respond more smoothly when "unexpected" phenomena occur.

9.3.2.4 Know the Capabilities and Limitations of Current Flood Protection Measures

It is desirable to comprehensively consider the structure of flood control facilities that can mitigate damage caused by external forces in the event of floods exceeding the planned flood magnitude, based on a comprehensive judgment of factors such as ease of restoration in the event of damage, initial investment and restoration costs, social and economic impact in the event of damage, and maintenance costs. In particular, restoring a structure to the way it was before the damage without examining the mechanism of damage occurrence and strength should be avoided unless there is a special reason to do so.

9.3.2.5 Know the Functions and Limitations of Facilities around Rivers during Floods

In floods caused by rainfall in excess of the planned-for rainfall, bridges, roads, embankments, and other structures around rivers may suffer flood damage. Therefore, when constructing roads and bridges around rivers, it is advisable to know the river maintenance plan, to know the limits of river control based on the current maintenance status, and to consider structures and advance restoration methods in case of floods exceeding the design's tolerance.

9.3.2.6 Recognize the Importance of Facilities around Rivers during Floods

Among bridges, embankments, and other facilities affected by floods that exceed the base high water, those facilities that are considered important for evacuation and restoration planning should be considered for structural strengthening and other drastic measures.

9.4 Recommendations on External Forces That Caused the Disaster

9.4.1 Rainfall

When rainfall exceeding probability numbers due to a super typhoon is expected, sediment disasters and floods are likely to occur. In addition, the type of sediment disasters may differ depending on the cumulative rainfall, hourly rainfall, and geology. On the other hand, the sediment disasters that occurred in Fukuchiyama City, Kyoto Prefecture, Tamba City, Hyogo Prefecture, and Hiroshima City in August 2014 were caused by strong localized short-duration rainfall. For this type of rainfall, it is extremely difficult to predict rainfall, and the observation system to capture localized rainfall is inadequate. Based on the above, the following recommendations are made to the government and facility managers.

9.4.1.1 Consider Disaster-Prevention Measures in Response to Rainfall Patterns

For rainfall caused by typhoons and rainy fronts, it is important to prepare for the occurrence of disasters by drawing up a timeline (disaster-prevention action plan) in advance, such as what kind of disaster will occur and when to evacuate, by referring to rainfall observations over a wide area, rainfall forecasts, and typhoon path forecasts. On the other hand, for rainfall that is locally heavy for short periods of time (rainfall that is difficult to predict), it is desirable to strengthen the system for rainfall observation and public information, present the current status of observation, and educate residents so that they can voluntarily take disaster-prevention actions. In addition, it is desirable to develop a prediction method for the time when a large-scale collapse does occur, based on the knowledge of the relationship between rainfall and the possibility of collapse and past damages, as well as on the establishment of a monitoring system.

9.4.2 River Water Level

The rivers in the southern Kii Peninsula are influenced by the topography, meandering through valleys and valley-floor plains flanked by steep slopes. The river width is narrow compared with the basin area, which during precipitation events results in a rapid rise in water level and an increase in flow velocity. The river disasters caused by Severe Tropical Storm Talas were the result of external forces that far exceeded the external forces assumed in the design of the structures and the overflow of levees due to the torrential rainfall, which, while still a reference value, has a recurrence period of 200 to 5,000 years based on 72-hour rainfall. The flood control projects that have been carried out so far are considered to have been very effective in reducing damage caused by "unexpected" rainfall such as S.T.S. Talas. However, looking at the changes in rainfall patterns in recent years, it is highly likely that similar rainfall will occur in the near future, and it is important to be prepared for such rainfall.

9.4.2.1 Analyze the Situation, and Build an Evacuation System Based on
*   Rainfall Data over a Wider Area*

It is desirable to construct and improve a wide-area flood forecasting model based on meteorological observation, forecast data, and local monitoring data, which are currently under study. It is also desirable to establish a system for information dissemination and evacuation using this system.

9.5 Recommendations for Disaster Emergency Response and Recovery Activities

Immediately after the Kii Peninsula Flood Disaster, the national and local governments began to assess the situation in the disaster area, rescue victims, and provide relief. The Joint Survey Team on Geo-hazards in the Kii Peninsula Caused by Severe Tropical Storm Talas in 2011 also conducted a comprehensive survey of the affected areas in Nara, Wakayama, and Mie prefectures immediately after the disaster occurred to elucidate the damage situation and damage generation mechanism. This section summarizes the lessons learned from the emergency response and recovery activities immediately after the Kii Peninsula Disasters.

9.5.1 Establish a System That Allows Both the Public and Private Sectors to
*   Respond Immediately to Assess the Actual Damage in the Event of a*
*   Large-Scale Disaster*

Immediately after a disaster strikes, it is important to confirm the extent of damage. It is only natural that the national and local governments should immediately begin assessing the damage and responding to rescue and relief efforts for disaster victims, and in response to this, it is necessary to establish a system that enables private construction-related industries (especially those involved in geological survey, design, and surveying) to voluntarily conduct damage assessments. Many disaster-prevention agreements have been concluded between the national and local governments and various industries. However, most of the activities of the private sector were initiated by requests from the public to the private sector. It is important to establish a system in which private companies of construction-related industries can work together to initiate emergency responses on their own. In the Kinki Region, geological surveyors' associations, construction consultants' associations, and surveying and design associations have signed an agreement for cooperation in the event of a disaster, and they are attempting to cooperate not only in emergency response, but also in disaster drills during normal times.

9.5.2 Establish a Disaster-Related Information Database

It is desirable to establish a system (disaster-related information database) that allows related organizations, both public and private, to share the results of emergency damage surveys and information necessary for emergency response on the internet. If it becomes possible to sequentially collect not only information obtained by each organization but also information from related persons in the disaster area and use it for various purposes, it will prevent information confusion during emergency response and lead to effective use of information.

9.5.3 Utilize the Experiences and Information of Those Who Have Responded to Disasters

Many people engaged in emergency response have experiences and information that can be applied to disaster management activities. It is necessary to record their difficult experiences and challenges so that they can be used in the event of similar disasters.

9.5.4 Organize a Manual for the Findings Related to Geological Investigations Accumulated at Large-Scale Slope Failure and Debris Flow Sites

In geological investigations at sites where large-scale slope failures and debris flows have occurred, there have been efforts and innovations that are different from those used in ordinary investigations. Based on this experience, it is desirable to prepare a manual for geological investigations.

9.6 Proposals for Emergency Response Technologies

The floods and sediment disasters that have occurred throughout the Kii Peninsula are by no means exceptional. They have been occurring intermittently and regularly since the formation of the Kii Peninsula and are expected to continue. Based on the results of previous surveys and research, we have compiled recommendations for activities to prevent or reduce human and property damage caused by heavy rainfall.

9.6.1 Record and Preserve the Location and Magnitude of the Disasters

The locations of the floods and sediment disasters caused by Severe Tropical Storm Talas in 2011 were often in areas where similar damage has occurred in the past or in close proximity. It is advisable to record the location and extent of damage. This is considered to be effective in identifying hazardous areas in the event of heavy rainfall. In addition, the accumulation of disaster records is expected to contribute to the formulation of safer evacuation action plans for residents by allowing them to envision disaster scenarios.

9.6.2 Establish a Restoration Plan for the Facility Based on the Actual Disaster Conditions

It should be reiterated that structures intended for disaster prevention and mitigation have limitations in their effectiveness according to their design performance, and when the expected external forces are reviewed, the performance of the structure and its effectiveness in preventing disasters should be reevaluated. When restoring a structure after it has been damaged, it is desirable not to restore the structure in its original form as a general rule, but to examine the cause of the damage, review the external forces, and design the structure to ensure the necessary performance, or to examine the degree of destruction caused by external forces before making an overall plan to reduce the overall effect of such forces. In addition, since the time for disaster assessment is limited, it will be necessary to organize in advance the restoration method in case of damage.

9.6.3 Evacuate to an Appropriate Evacuation Site in a Timely Manner

When a disaster is expected, evacuation advisories should be issued early and with time to spare. In this disaster, there were cases in which residents did not evacuate properly due to past experiences of heavy rains that did not lead to disasters. It is also important to ensure appropriate evacuation sites for different types of disasters. In mountainous areas in particular, it is difficult to secure an evacuation site, but an emergency evacuation site may be sought in a road tunnel. In this case, it is necessary to take measures to prevent accidents between vehicles and evacuated residents. Photo 9.1 shows a cenotaph erected in the Ui Area of Gojo City, Nara Prefecture, where a large-scale slope failure occurred. Unfortunately, 11 people were killed by the collapsed sediments in this area. Since it was impossible to escape after the collapse due to the suddenness of the large-scale slope failure, this cenotaph tells us that early evacuation is very important.

9.6.4 Conduct Disaster Drills That Sustain Residents' Interest

It is desirable to continuously promote disaster-prevention education for men and women of all ages by holding many events such as disaster drills that are friendly to local residents and games that incorporate elements of disaster-prevention training while having fun.

Photo 9.1 Cenotaph erected in the Ui area where the site of a large-scale slope collapse is located (Photo by Ryoichi Fukagawa).

9.7 Proposals for the Preservation of Cultural Heritage

The Kii Peninsula is home to the World Heritage "Sacred Sites and Pilgrimage Routes in the Kii Mountain Range." This cultural heritage site was also damaged by Severe Tropical Storm Talas in 2011. In addition, there is no law that covers the protection of World Heritage sites, but they are managed by multiple laws and institutions. Temples and shrines are protected by applying the Law for Protection of Cultural Properties by registering them as cultural property buildings along with the core zone of the pilgrimage route, which is directly designated as a historic site. A buffer zone, which is a restricted-use area around the core zone, is protected by applying the Natural Parks Law, River Law, and ordinances of related municipalities. In addition, the famous Kumano Kodō, a pilgrimage route to the three mountains of Kumano, often passes through areas prone to sediment disasters and serves not only as a historic site but also as a road for the general public, requiring prompt disaster recovery. Based on the above findings, we have compiled recommendations for the preservation of cultural properties from natural disasters.

9.7.1 Develop a Centralized Management System and Legislation for the Conservation of World Heritage Sites

The Kii Peninsula has existed as a place of faith since ancient times. In particular, the cultural heritage centering on the Kumano Kodō Trail is a World Heritage site. It is desirable to establish a centralized management system for the conservation (including disaster prevention and risk management) and utilization (use by tourists) of World Heritage sites, and to enact laws to protect World Heritage sites directly.

9.7.2 Form a Common Understanding of Cultural Heritage Disaster Prevention

When there are multiple jurisdictions involved in cultural heritage, it is necessary to share the recognition that it is cultural heritage from ordinary times, and each stakeholder should have knowledge of cultural heritage conservation.

9.7.3 Improve the Efficiency of Cultural Heritage Maintenance Work

Cultural heritage preservation requires routine management, and regular monitoring is desirable. In addition, if a large amount of labor is required, it is necessary to consider some way to reduce the amount of labor.

9.7.4 Consider Crisis Management for Important Cultural Heritage

The Kumano Pilgrimage Route (Yokogaki Pass) and Nachi Great Falls are cultural heritage sites with natural scenery and nature itself, and it is difficult to take disaster-prevention measures even if hazards can be identified in advance. Therefore, it is important for crisis management to predict the damage before a disaster occurs, envision disaster scenarios, understand the relevant authorities in such cases, and consider how to cope with the aftermath. For restoration, it is necessary to consider how to preserve cultural heritage in a resilient manner from a long-term perspective, bearing in mind that cultural heritage is an asset that will be handed down 100 to 1,000 years from now.

9.7.5 Consider Cultural Heritage Disaster Prevention Multilateraly

In cultural heritage disaster prevention, it is necessary to carefully consider and plan for (A) preservation of the cultural heritage itself, (B) restoration after the disaster, and (C) safety of people surrounding the cultural heritage (tourists, tourism industry workers, local residents, and so on).

9.7.6 Improve the Level of Disaster Preparedness by Use of Monitoring Technology

For cultural heritage sites where reoccurrence of disasters is predicted, it is desirable to improve the level of disaster preparedness by conducting necessary monitoring.

9.7.7 Create a Social Consensus That Cultural Heritage That Cannot Be Fully Restored Is also Valuable

Slopes deteriorate over time due to weathering and other factors, and eventually collapse and flatten, making it difficult to preserve the path through them over the long term. It is hoped that a social consensus will be formed to recognize the value of this special form of cultural heritage not only in terms of the material value of the original path that remains intact from ancient times, but also in terms of its spiritual value, and that the value of a new path with a similar landscape using similar cobblestones will be recognized when it is established.

9.8 Recommendations for Local Residents

The large-scale disaster caused by S.T.S. Talas in 2011 brought damage to a wide area of the Kii Peninsula. Such damage has been repeated many times in the past, and the wisdom to cope with disasters has been passed down considerably. However, some people said, "I never thought my house would be damaged." Residents' awareness of disaster prevention is not necessarily high, and their disaster-prevention activities during normal times are far from perfect. We have made our recommendations in order to raise the residents' awareness of disaster preparedness. Since the affected area has a large proportion of elderly people, the recommendations were written in as plain a language as possible.

9.8.1 Recognize That Disasters Can Occur Anywhere

Make residents aware that large-scale slope failures, debris flows, floods, and other disasters such as those that occurred on the Kii Peninsula can occur anywhere in Japan, which has a lot of steep mountainous terrain (and perhaps in other countries around the world with similar conditions). Unlike earthquake disasters, sediment disasters do not strike suddenly, so when the weather forecast indicates heavy rain, have the participants consider whether they need to evacuate and whether they will be able to do so in a timely manner. Since the amount of rainfall that can cause a disaster differs greatly from region to region, we encourage people to pay attention to information provided by local authorities and consider taking action (evacuation) based on their own judgment as soon as possible. Local governments do not have a complete picture of everyone's situation and respond to disasters based on a variety of information. Therefore, they should be aware that the larger the disaster, the more difficult it becomes to consolidate information and the less detailed the response will be.

9.8.2 Prepare for Disasters on a Daily Basis

It is important to confirm evacuation sites and routes together with neighbors. Since the degree of danger differs between evacuation during daytime and evacuation during nighttime, publicize the fact that evacuation at night is particularly dangerous and that people should take that danger into consideration. Also, have them confirm the evacuation site and evacuation route.

9.8.3 Enhance Local Disaster Preparedness

The national "Guidelines for District Disaster Prevention Plans" point out the importance of district disaster prevention promoted by district residents and businesses, rather than local disaster prevention promoted by local governments, in order to improve disaster preparedness. Immediately after a disaster strikes, self-help and mutual aid by district residents are very important, and district residents are encouraged to work together to reduce the damage. In addition, in order to build a community that is resilient to disasters, it is important for local people to work together. Local residents themselves should create their own evacuation plans tailored to their actual situation, which will increase the community's ability to prevent disasters. Participation in disaster drills and other events will strengthen community cooperation.

9.8.4 Communicate the Experience of the Disaster

In communities hit by disasters, disaster drills and other opportunities should be used to pass on the experiences of the disaster and to encourage the government and facility managers to share their experiences with a large number of people. Photo 9.2 [1] was taken at a memorial service held in front of the cenotaph in Ryujin Village, Tanabe City, Wakayama Prefecture. In the Shimoyanase Area of Ryujin Village, a large-scale slope failure occurred

Photo 9.2 Memorial service held in front of the cenotaph in Ryujin Village, Tanabe City, Wakayama Prefecture (Photo by Seisuke Ushiro).

during the 1889 Meiji Totsukawa Disaster, and a natural dam was formed with the collapsed sediment. When that natural dam broke, as many as 83 people in this area were killed. It was a memorial service 130 years later, and the memory of the flood damage will be passed on to the younger generation.

9.8.5 *Consider Disasters as Tourism Resources*

Disasters can cause tremendous damage to people, but we should also think about using disasters to local advantage. The Kii Peninsula is a region that has suffered from frequent disasters, but as a result, it is also a region rich in tourism resources. In order to develop the local tourism industry, it is important to have an evacuation plan that cooperates not only with local residents but also with tourists, which is also necessary to promote tourism. Disaster-resistant tourist destinations have strong appeal, and it is worth considering the use of disaster-prevention resources for tourism.

Finally, the following five articles on disaster prevention (for sediment disasters and floods) are listed below as recommendations for local residents.

1 When there is heavy rain, think that sediment disasters and floods are likely to occur.
2 When a disaster is likely to occur, cooperate with local authorities and neighbors to evacuate as soon as possible.
3 Cherish the wisdom of disaster prevention that has been passed down from generation to generation in your community.
4 Be sure to pass on your experiences of disasters to your children, grandchildren, and the next generation.
5 We never know when and where a disaster will strike. We should be vigilant on a daily basis.

9.9 Summary

This chapter summarizes recommendations for governments, facility managers, and the general public in the hope that damage from large-scale heavy rainfall disasters expected to occur in the future will be reduced.

With regard to disaster forms such as sediment and flood disasters, we pointed out the importance of understanding the relationship among topography, geological conditions, and rainfall conditions that affect the scale of the disaster, as well as the area's disaster history.

The importance of a timeline approach with regard to external forces that caused the disaster, such as rainfall and river levels, and the need to establish a rapid evacuation system based on the results of on-site monitoring and so forth, were also indicated.

With regard to emergency response and recovery activities, we pointed out the importance of establishing a system for promptly confirming the extent of damage and building a database of disaster-related information.

With regard to the preservation of cultural heritage, we pointed out that it is important to have a common understanding that cultural heritage is public property, and that it is important to form a centralized management system and rules on this basis.

Finally, recommendations for the general public were summarized in the Five Articles of Disaster Prevention (sediment and flood disasters).

Reference

[1] Ushiro, S.: Kii Peninsula Great Storm, *Haru Shobo*, 2022. 8. (in Japanese)

Index